SHARP

SHARP

50 Simple Ways to Improve
Your Life with Brain Science

THERESE HUSTON, PH.D.

Originally published by Mayo Clinic Press, 200 First St. SW, Rochester, MN 55905

First published in Great Britain in 2025 by Short Books, an imprint of
Octopus Publishing Group Ltd
Carmelite House
50 Victoria Embankment
London EC4Y 0DZ
www.octopusbooks.co.uk

An Hachette UK Company
www.hachette.co.uk

The authorized representative in the EEA is Hachette Ireland, 8 Castlecourt Centre,
Dublin 15, D15 XTP3, Ireland (email: info@hbgi.ie)

ISBN: 9781804193402
eISBN: 9781804193426

A CIP catalogue record for this book is available from the British Library.

Printed and bound in Great Britain

10 9 8 7 6 5 4 3 2 1

The publisher wishes to thank John R. Henley, Ph.D., M.S.

This FSC® label means that materials used for the product have been responsibly
sourced.

The medical information in this book is true and complete to the best of our knowledge.
This book is intended only as an informative guide for those wishing to learn more
about health issues. It is not intended to replace, countermand or conflict with advice
given to you by your own physician. The ultimate decision concerning your care should
be made between you and your doctor. Information in this book is offered with no
guarantees. The author and publisher disclaim all liability in connection with the use of
this book. The views expressed are the author's personal views, and do not necessarily
reflect the policy or position of Mayo Clinic.

**Proceeds from the sale of every book benefit important medical research and
education at Mayo Clinic.**

*To my husband, Jonathan, because geeking
out together is worlds more fun*

Contents

Be Better to Others

Master Your Destiny

Relax and Connect

Introduction

Let's talk about the unpredictability of your days. See if this sounds familiar: Sometimes you feel witty, sharp, and on top of things. Ideas just *flow*. But other days, you struggle to keep up, you're mentally foggy and a little slow on the uptake. You find yourself wishing that people would repeat themselves. It doesn't matter how smart you are or how big your paycheck is; we all have those days.

Or consider your productivity: Some days your work glides seamlessly from the moment you sit down at your keyboard, fingers flying, priorities crystal clear, phone (mostly) easy to ignore. Other days you just can't get started and you're not sure why. You feel blunt, ineffective. You watch the hours disappearing, but you're stuck and just can't get motivated.

You feel the difference in your relationships, too. Sometimes you and your partner are in sync, and you know exactly what to do when your partner is stressed. Other times, no matter what you say, it's the wrong thing, and you find yourself tiptoeing, wishing you could find the right words.

You don't need to be at your best every single day, thank goodness, but it sure would be nice if being high functioning could feel, to some extent, more within your control.

With this book, it can be. In these pages, I will give you tools so that you can be "on" when you most need to be. At work, at home, and everywhere in between.

Your brain happens to be the most powerful tool you'll ever own. It's with you everywhere you go, but unfortunately, it didn't come with a user's guide.

Until now.

Say hello to the book that will help you make the most of the brain you've got.

The User's Guide You've Been Missing

The way I see it, there are basically two ways to think about your brain:

1. It serves you incredibly well; or
2. It could serve you better. (In some cases, that might feel like an understatement.)

Most people I know, myself included, are in the latter camp. We wish our brains worked better. Some of my younger female friends complain about "pregnancy brain"—the memory loss and brain fog that many women experience before their babies are born.[1] My older friends, regardless of gender, shake their heads and bemoan the fact that they're just not as mentally sharp as they used to be. All too often, they say, they'll walk purposively into a room and then stop, thinking, *Wait, why am I here again?*

Even people who don't have maternity or age working against them still have their sacred little wish list when it comes to how their brains function. They wish they could focus better, make tough decisions with more confidence, or feel less aggravated by life's little assaults. If you're seventeen or eighteen years old, you might feel like you could solve the world's problems if someone just gave you a chance. But the rest of us? We've at least been given a chance to solve some kind of problem and we're doing okay, thankfully, but deep down we know we could do better. We've got this nagging feeling that something is *missing,* that our truest potential hasn't been unlocked.

I'm here to help you unlock your full potential.

Or, if I'm a little more honest and a lot more modest, neuroscience is here to unlock your full potential, and I'm here to be your guide. I'll wade through the research, translate the most promising findings, and explain it in ways you will not only understand, but about which you can get excited.

Some of you are probably armchair neuroscientists and will want all the specific details I could provide and more. You are my people.

But I also know what's it's like to have your eyes glaze over when a scientist geeks out. That's why I'll strive to translate the science clearly so that you feel you can use it immediately, and I'll try to keep my geek-outs to a minimum. (But the thing is: I believe that becoming smarter and more productive by simply investing 5 to 20 minutes of your day is something to seriously geek out about!)

I'll also teach you more about how your brain works. I find almost everyone perks up a little when they learn some juicy bit of brain science, as long as the information is clear, approachable, and useful.

What perhaps excites me most is that you can do something no neurosurgeon, no matter how talented, can currently do. You can give yourself a better brain. You can actually *add* gray matter with some of the practices in this book.

Throughout this book, I'll be translating rigorous research into practical strategies, but when you're feeling skeptical (and chances are you will from time to time), I'll do my best to persuade you. I won't cheerlead—no miniskirts or megaphones—but I can, and will, empathize. I too felt skeptical several times as I read some of the strategies I'm about to describe. But I nonetheless tried those approaches that felt just a little too weird to work. And if they're included in this book, it means I found they made a difference.

If You're Looking for Neuroscience 101...

I like truth in advertising, so just to be clear, this isn't Neuro-science 101. Maybe you're going to stop reading now, and that's fine, no hurt feelings.

What's the difference? First and foremost, in a Neuroscience 101 course, you'd learn the names of at least 50 brain areas. (Trust me, I've taught Introduction to Cognitive Neuroscience, and it takes *many* flash cards to keep everything straight.) Here, though, I'm going to focus on about a dozen areas, those fascinating brain regions that are crucial to being brilliant and effective *and* to being a caring, decent human being in life in general. If you know about these brain areas, you'll not only feel sharper but also be able to talk intelligently about how you learn, remember, make decisions, pay attention, show empathy, avoid bias, and cope with stress.

I'll also be simplifying the story. Whereas a Neuroscience 101 course would offer a *War and Peace* version of the different brain areas, I'll basically be giving you the CliffsNotes. Take anxiety, for example. When you feel anxious and strive to cope with that anxiety, there are at least eleven different brain areas involved in that process.[2] Describing them becomes pages and pages to wade through (and reading it might inadvertently stimulate those anxiety-sensitive regions). To simplify each chapter and topic, I'll zero in on one or two brain areas that play a particularly important role. I'll aim to keep the neuroscience short, sweet, and to the point, and I promise not to overload you with terms you'll wonder why you ever learned.

In this book, you'll also be getting something that you don't get in most lecture halls. Toolkits. At the end of each chapter, you'll have a fun "Try this" list consisting of practical strategies that have been given the thumbs-up by scientists. Most of these strategies are free, but if you like using apps to guide your progress, I'll describe what to look for in an app and what to avoid. My goal here is to offer you serious science, filtered for practicality.

As you'll soon see, some chapters have several strategies (for example, there are five strategies for improving your creativity in chapter 2) whereas others only have a few (there are only two strategies for feeling more empathy in chapter 8). It's not that empathy is less important than creativity. Rather, it's that empathy is a more nuanced topic and there are several popular misunderstandings about it, so I'll spend a little more time distinguishing what works from what doesn't and why. I want you to be smarter than the average bear, and I don't want you to pursue an unproductive strategy or naively assume that any hack that's trending is the way to go.

I've also included two things that I love in this book that I wouldn't put in most university courses. First, I'm all about myth-busting, so several chapters have a "What Doesn't Work" section in which I dispel a popular strategy that, frankly, science doesn't support. Second, I want this book to help as many people as possible, so some chapters have a quick sidebar titled "Does Age Matter?" When I find solid science revealing a better strategy for adults over age sixty, I'll include it.

Finally, there's one last important way this book differs from Neuroscience 101. In a university course, the professor would assume that you're devoting several hours to reading and studying the material every single week. That is not an assumption I will make. If anything, I'm hoping you'll take 15 minutes to do a little research of your own, but it's not necessary. I'm assuming that you're squeezing this book into your very busy lifestyle, reading bits on the bus, over lunch, right before bed, whenever you can, hoping to find new tactics you can fit into your schedule. With the busiest version of you in mind, I've also tried to provide at least one strategy in every chapter that takes 10 minutes a day or less to implement. There are plenty. If you find yourself with even less time, check out the Appendix at the back of the book. It tells you where to find the quickest strategies.

Are We There Yet?

Every book has its origin story. This one started over thirty years ago.

When I began graduate school in the early 1990s, my goal was to apply neuroscience to real-world problems. Specifically,

I wanted to know whether, with practice, people could become more skilled at multitasking. (The science examining multitasking was very new back then, and we didn't even call it that—we called it "task switching," a much less sexy phrase, but ultimately, a more accurate one.) I also wanted to know which areas of the brain were involved in multitasking. If someone became better at multitasking, were they drawing upon different brain regions or were they using the same ones they always had, just more efficiently?

The science wasn't there yet. Not even close. As a grad student, I would sit at the back of a dark classroom, hands wrapped around a cup of instant General Foods International Coffee to keep me warm and awake, while a researcher clicked through their monotonous PowerPoint slides of what was evidently cutting-edge research. I remember one scientist describing, in a particularly dull presentation, how his team had flashed the image of a black-and-white checkerboard repeatedly at test subjects to see how their visual cortices, the part of the brain responsible for vision, would light up. Flashing black-and-white checkerboards while someone passively lies in a brain scanner, daydreaming, was considered state-of-the art research? Why weren't scientists asking the important, practical questions that could help people improve their lives?

But back then, neuroscience simply wasn't ready for practical applications. The most practical discoveries of that era, such as how to use cognitive behavioral therapy to help people cope with trauma, were being made by traditional psychologists who were using well-established research methods. But the discoveries being made by psychologists using neuroimaging? Their insights were more like forehead slaps. These scientists were coming to brilliant conclusions like, "We need to tell women

not to wear underwire bras into a room with a giant magnet."
(I wish I was kidding.)

I think of it this way: Neuroscientists who were figuring out functional neuroimaging technology in the early 1990s were somewhat akin to the Wright brothers establishing airplane technology in the early 1900s. They were innovating, of course, but they were taking baby steps. They needed to prove you could trust the technology, that it was not only safe but highly reliable, and they needed to show that you could achieve the same fundamental results not once or twice but a hundred times.

But I wasn't interested in the fundamentals. Whereas most neuroscientists were just making sure they could get the plane off the ground, I was asking how many bags I could check. I'm not saying I was precocious. Hardly. Just practical and, well, a tad impatient.

So I focused on teaching instead, where I discovered that I loved making hard concepts easier to understand. I became an assistant professor, I taught neuroscience courses, and although I dabbled in some testosterone research, I largely set aside my desire to do practical neuroscience research. About a decade later, in the early 2000s, I returned to it. At that point, I had become passionate about education. Some of my colleagues and I were brainstorming a grant proposal for the National Science Foundation about finding the best way to apply neuroscience to everyday classroom education. We realized that our proposal needed more gravitas, so I approached a prominent neuroscientist I knew, one of the best in his field, and asked if he'd like to join us. His response was essentially, "Applying neuroscience to education? Too soon. We don't know enough yet."

Again, I set aside my neuroscience interests for more practical pursuits and focused on helping good professors become students' *favorite* professors. I loved my work. I published a few books applying traditional psychology research to everyday problems and largely forgot about my neuroscience ambitions.

And then one day in early 2021, I stumbled upon the Huberman Lab podcast. Dr. Andrew Huberman is an associate professor of neurobiology at Stanford who takes complex, high-level concepts in neuroscience and translates them to everyday use, which is exactly what I'd always been looking for. As I listened to him apply neuroscience to everyday challenges, I realized the science had finally arrived. (I want to acknowledge that Huberman has been criticized for certain issues, some personal and some professional, but the discovery of his podcast remains a lightbulb moment for me.) I dove headfirst into the research literature to catch up, and this book is the culmination of the best of that research.

Becoming Sharper and Taking Shortcuts

As you read this book and start using some of the strategies we discuss, you're bound to become a sharper version of yourself. Merriam-Webster's Dictionary defines the word "sharp" as "keen in intellect, perception, and attention." That's what I want for you. On days you're feeling sharp, you think more quickly, and you solve problems more easily. You focus like a concert pianist. You're not dull or slow when you're sharp, but you're also not thrashing about, fruitlessly trying things that don't work. You're the most effective version of yourself. With the help of this book, you'll not only recognize those moments when you're not living up to your fullest potential but you'll know how to dial up specific abilities so you can perform at your best.

This book is a valuable resource, yes, but your principal resource—already at your disposal—is, quite simply put, your brain. And realizing that you have the resource you need most makes all the difference. As you'll see in chapter 13, a stressful situation feels more threatening when you don't think you have sufficient resources to cope. A hard feedback meeting with your manager, for instance, or a presentation in front of a roomful of people, can be incredibly intimidating when you feel you don't have what you need to succeed.

It probably feels like the resources you need in that moment are more time or support (or maybe a crash course in public speaking?), and sure, they could help, but the amount of time and support you receive is often outside your control. What is largely within your control and at your disposal? Your brilliant brain. As you read this book, you'll become more confident and less stressed because you'll know exactly what to try before you walk into that meeting or up to that podium. Obviously, you use your brain all the time, but compared to how you could use it, chances are it's an underutilized resource.

Let's clarify something up front. I'm not claiming that you only use 10% of your brain. That's a very popular misconception—more than 1 out of every 3 individuals worldwide believes that misleading notion is true—but it's completely inaccurate.[3] As one neuroscientist at Mayo Clinic observed, "Evidence would show that over a day you use 100% of your brain."[4] Even when you're simply relaxing and letting your mind drift, you use much more than 10%, and when you're solving hard problems, neuroimaging would reveal that *many* brain regions fire up. We also know we use our whole brains because when people have damage to even a small region of their brain, they often have debilitating impairments, and that wouldn't happen if 90% of your brain was just sitting around, underutilized. (If your mum, for instance, suddenly showed signs that she was having a stroke at dinner, you wouldn't shrug and say, "It's okay, Mum. Finish eating your lasagna. If you are having a stroke, chances are it will be in the 90% of your brain that you don't use.")

So when I say that this book will help you use your brain to its fullest potential, I mean you'll be tapping into areas that you already use but could use *more* efficiently or effectively. You can be sharper. I love a good analogy, and I think a driving one works here: Basically, I'm going to teach you how to take short-cuts. Imagine that you're driving to a destination you already know. You get in your car and go. You operate intuitively and on autopilot. You take the most familiar route, the one that is routine. Yet one day, there's construction, and you curse how ridiculously long that route, one that normally works, is taking you.

With a little prep work, however, you could have avoided that frustration. If you had checked your favorite driving or mapping app before leaving home, say, you might have discovered there was a faster, more pleasant way.

Likewise, when you're trying to reach a solution at work or at home, you'll often do what you've tried before. And it works. Sort of. You activate the same brain regions in the same ways you always do, completely unaware that those areas may already be overloaded, and that there might be a better way.

Imagine, for example, that you're about to go to an important meeting where you'll be in a room full of new people whose names and faces you're going to want to remember. You could use whatever intuitive memory strategy has worked for you in the past, even though historically you've only been able to memorize the names of three or four people before they all begin to blur. And that's when you're relaxed! You're far from relaxed right now, though, because you'll be giving a presentation to this group, and your brain is in a highly overloaded, congested state. But you don't know another memory strategy, so you cross your fingers and use the same tactic you've always employed.

I can offer some surprising tools to help you remember those names (chapter 6) and deal with the stress you're feeling (chapter 13) based on the brain science of memory and what we know about strategies for reducing stress hormones. Once you know these strategies, you'll find it easier to both remember the names *and* manage your stress. Predictability and improved performance when you need them most? Check. Plus, you'll have some new tools you can use even when your stress isn't high. You'll become a little sharper with each and every chapter.

There will be times when I offer a strategy that is backed by solid psychology research, so we know it works, but neuroscientists haven't pinpointed *how* it works in the brain just yet. I include these gems because they can help you, the social science is solid, and I believe the brain science will soon follow.

The Question on Everyone's Mind

As you read each chapter, your first question will probably be whether you have to *exactly* follow the protocol I'm describing. For instance, if the advice is to meditate for 13 minutes, would 10 minutes be enough? How about 5 minutes? Will we get the same benefits? If being sharper is all about taking shortcuts, you might find yourself asking, "Is there an even *shorter* shortcut?"

I often found myself wondering about these kinds of things, and whenever possible, I'll provide information about other protocols researchers have tested. But the reality is that most researchers don't try several variations. They want to see significant results, and if they aren't expecting there to be a huge difference between 10 minutes of an activity versus 13 minutes, they won't waste their time and resources testing that difference.

My suggestion would be to aim for the suggested protocol but don't beat yourself up if you don't follow it to a T. The goal here is progress, not perfection. Take meditation, for example, which we'll discuss later. If you go from hardly ever meditating to meditating 5 minutes a day and you can keep that up for three weeks, I suspect you'll see at least *some* of the promised benefits. Plus you'll learn a lot about yourself in the process. You'll learn, for instance, whether a meditation practice is something you want to make more room for in your life or if your time is better spent on other changes.

I hope that your mindset as you read this book will be that you're experimenting with yourself. Try something and pay attention to whether you see a change you like. I've tested almost every single strategy I'm going to present to you, and over the course of a year, I've become more resilient in difficult situations, more focused and productive, and *a lot* better at managing stress. I went from sleeping 7 hours a night to almost 8 hours, largely because my stress levels are lower. I spend less time working and accomplish just as much.

It's important to clarify that I'm a scientist, however, not a medical professional, and the information provided in this book is not medical advice. It's a resource for educational purposes. Everyone's individual health needs are unique, and it's crucial to consult with a qualified healthcare professional before you make significant lifestyle changes. I'll offer research, inspiration, guidance, and, as promised, a little persuasion, but please, prioritize your health and safety by seeking professional medical advice when needed.

And just in case you're wondering, I have no financial relationships to any of the apps or supplements I recommend. No one has asked me to promote them. I stumbled upon these through my own rigorous research, and I hope they help you the same way they've helped many others.

Let the Fun Begin!

This book is organized into two parts. The first part is about being mentally sharper at work. Nine chapters will take you through everything from starting a project to handling a Q&A when you present that project. The second part of the book applies similar principles to your personal life. That's where you'll learn how to tackle big challenges such as making better life decisions, managing chronic stress, improving your health, and supporting your partner. (I thought about addressing sleep, but Matthew Walker's book *Why We Sleep* is so thorough, I wasn't sure what I could add. I strive to write what most needs to be written.)

Although I've organized the book into the personal and the professional, most of the strategies you're about to learn can be used for *any* part of your life. For example, the chapter on getting motivated is in the section on thriving at work. But if you're like me and dread filing your income taxes every year and put it off longer than you should, please take the wisdom from that chapter and apply it to organizing your home and personal life.

This book is a true "choose your own adventure." Read it cover-to-cover or jump around, picking whichever topics will help you most right now. There are some concepts (like the function of dopamine) and practices (like exercise) that are talked about in multiple chapters, so I've indicated which other chapters to check out if you want to learn more. If there's ever a strategy that feels too woo-woo for you, skip ahead to the next section and find something that's more your speed. Make this book your own.

My hope is that if I ever happen to meet you and you show me your copy of this book, you'll have filled it with colorful Post-it notes or you'll have dog-eared so many pages that it's twice its original thickness. I can honestly say that I had fun writing this for you and I hope you'll have even more fun changing your life with it.

Let's go discover the keenest, sharpest version of you!

THRIVE

9-to-5

Let's start by improving your work life. On average, most of us spend one-third of our adult lives at work, potentially doing tasks we didn't choose or don't feel we do well. At least, not yet.

Small changes grounded in neuroscience can make hard days much easier, and they just might turn good days into phenomenally productive ones. We'll start with some tools for getting started on a project, then we'll move to approaches for doing your best work, and we'll wrap up by looking at how you can be a better colleague.

GET STARTED

CHAPTER 1

Get Focused

It's Monday. Worse yet, it's 3:45 p.m. on Monday, and you still haven't started on your most important project for the day. Where did the time go? You'd blocked off this afternoon to work on it, but you're finding it hard to focus. First your phone interrupted you, then your colleagues, then your manager, then your phone again. It's felt like a parade of distraction (but unfortunately, there weren't any clowns throwing candy).

Should you even try to work on that task this late or should you just leave it until tomorrow morning, when you can start fresh?

First of all, don't kick yourself, especially if your phone is your major distraction. Social media and marketing companies spend billions of dollars every year to figure out how to keep your eyeballs away from your work. Simply having good intentions to ignore your phone from nine to five is like bringing a knife to a gunfight. You're going to lose.

You need more than good intentions and an open block on your calendar to get your most important work done. You need strategies for focus.

As for whether you should wait until tomorrow, the first thing to ask yourself is this: Are you more of a morning person or a night owl? Yes, psychologists have found there really are two types of people, and if you're a true night owl, your mind is more likely to wander in the mornings.[1] If you're a night owl, keep your afternoons or evenings free for your most important focus time. If you're a morning person, block off the first hours of your day. But if you want to dial up your focus any time of day (or you simply don't have the scheduling luxury of choosing between the two, like many parents, say, whose evenings are chock-full of family duties), neuroscience has got your back.

WHAT DOESN'T WORK

The neuroscience is clear on one thing you shouldn't do: multitask. Don't start your most important work, then toggle back and forth between it and your email. It's tempting, but terrible. It might feel like you're accomplishing two things simultaneously—look at you go!—but the reality is that "multitasking" is a misnomer. You're actually just switching back and forth between tasks, and switching has a cost. A big one.

As I previously mentioned, at one time I wanted to study the neuroscience of multitasking in grad school. After conducting research on it, though, I quickly learned that not only is multitasking ineffective but it can actually damage the quality of your work. Neuroscientists have found that when people switch back and forth between their main task and a side task, they have reduced brain activity for the main task and, as a result, they start making more mistakes. *Many* more. In one classic study of multitasking, brain activity dropped by a whopping 37% for the main task while mistakes went up by 47%.[2] So instead of making five or six mistakes as you work through your most important task of the day (which is already, I'm sure, more than you want to make), you'll make eight or nine mistakes. It's a huge handicap. When you finish, you'll probably be frustrated that it's not your best work, and you'll likely have to spend more time correcting those mistakes later.

Worse yet, if you multitask a lot, you'll find it even harder to focus in general. A team at Stanford has found that individuals who admit that they do a lot of multitasking have more frequent lapses in attention, even when they're seemingly focused on one thing.[3] They blank out for a second or two in meetings and in conversations, like we all can do from time to time, but they do it a lot. And they're more forgetful than their peers, perhaps because their attention more frequently lapses at key moments. If you're someone who is constantly checking your phone or email, you're likely making it harder to concentrate when you really need to.

Admittedly, neuroscientists haven't yet determined which came first—do people who already struggle with frequent lapses in attention turn to multitasking because their attention won't stay put, or does multitasking cause more frequent lapses in attention? Time and more research will tell. But if you need laser-like focus, you probably don't want to be the guinea pig.

Perhaps you're thinking, *But I'm really good at multitasking, better than most people!* If that's the case, please try this quick test. Start a timer and see how long it takes you to recite the alphabet out loud from A to N. Check the number of seconds and jot it down. Start a timer again and see how long it takes you to count from 1 to 14 out loud. Write that down as well and add it to whatever you got for the alphabet. That's your total time.

Now start a timer and alternate between saying a letter and saying a number, going from A to N and 1 to 14. So it's A1, then B2, etc.

Go.

How did you do? If multitasking was as easy as doing the two tasks individually, then your total time for reciting the alphabet and counting should be the same as alternating between the two (A1, B2, etc.).

Chances are the two aren't even close. Even if you're "really good" at multitasking, it probably took you twice as long to alternate back and forth—for most people it takes at least three times longer, plus they make mistakes they don't make when they do each task separately.

So even if you *are* really good, you'd still be faster if you completed one task and then switched gears. Many of us think we're excellent at multitasking and have no idea how much slower we've become.[4]

You might be thinking, "I can't just ignore Slack and email. My manager might ask an important question and need an immediate reply." To sidestep this, you could preemptively post an auto-reply like, "Doing focus work, available again at 4:00" or ask your manager how they'd like to be notified when you're doing your most important focus work. Most managers will appreciate protected focus time, especially for work they too prioritize, as long as you're available at predictable times.

Perhaps you're like me and your greatest distraction is your computer. It's painful that the one tool that helps me the most also hurts me the most. If that's the case, then a focusing app might be your new best friend (or at least your new productivity buddy). A good focusing app will block websites and apps, has a timer or scheduling option, and should be a little cumbersome to disable. You can set it to run from 3:00 to 5:00 p.m., let's say, or whenever your best time might be. And, if you're someone who finds yourself being pulled back into your work at all hours, you could also use a focusing app to block Slack (or whatever the messaging app du jour might be) once your workday is done. Three of the top reviewed focusing apps that meet my criteria are Freedom, Cold Turkey, and Focus Bear. They take a little time to set up initially, but according to most users, the productivity you gain is well worth it.

What Works

1. Tetley over Costa: Drink Tea

Around 80% of the world's population consumes caffeine every day, and if you're among them, you already know that it perks you up.[5] Just one cup can snap you out of your morning fog or help you resist the gravitational pull of a post-lunch nap.

But not every cup of caffeine is created equal. When it comes to focus, tea is better than coffee (sorry, latte lovers!). Tea is more effective because it contains an amino acid called l-theanine. Any type of caffeine can improve attention and suppress sleepiness, but if you are really struggling with distractions, what you may need is the one-two punch of caffeine plus l-theanine. Researchers find that caffeine alone increases alertness but caffeine combined with l-theanine allows people to resist distractions and perform tasks faster.[6] It's you, but a more focused version of you.

Caffeine plus l-theanine is especially helpful when you need to toggle between tasks. I know I just said that multitasking is bad news, but the reality of our current work culture is that there are times when you simply can't avoid it. You may have to pivot from your most pressing task to answer two urgent emails, but if your body has enough l-theanine, you'll be able to dive back into your important work more quickly. (As you can see, l-theanine seems to do it all when it comes to boosting focus.)

There's another added benefit to l-theanine: It makes you calm. People often comment that when they drink a lot of coffee they feel jittery, but when they drink a lot of tea, they feel alert but smooth. That's the l-theanine. L-theanine reduces anxiety and stress levels.[7] If you have trouble falling asleep at night because your mind is racing, your doctor may recommend taking l-theanine before bed. But don't worry, taking a little l-theanine at 3:00 p.m. won't knock you out. It's not a sedative, it just produces calming effects on the brain and nervous system, counteracting some of the less pleasant side effects of caffeine.

How does l-theanine plus caffeine lead to better focus? Neuroscientists are still asking that question, but one promising finding is that this potent chemical combo reduces activity in what's called the "default mode network" of the brain.[8]

The default mode network is a network of roughly 10 different areas spread throughout the brain.[9] Neuroscientists noticed that when participants were in a brain scanner but didn't have anything to do or focus on, this combination of regions would consistently light up. It seems to be the pattern of activity that your brain defaults to when you're awake but waiting, much like when your car is idling. At first, researchers thought that when the default mode network lit up, it basically meant your brain wasn't doing much.

Since then, though, researchers have learned that this network tends to be highly active when your mind is wandering (which is why it lit up when people had nothing to do). Caffeine plus l-theanine reduces mind wandering. By reducing activity in the default mode network, caffeine plus l-theanine stops your brain from being drawn to pursue every new thought that pops into consciousness. It helps you focus on whatever you prioritize.

Different teas vary widely in how much l-theanine they contain, but black, green, and white tea all contain some.[10] (You might have heard that green tea contains more l-theanine than black, but lab testing reveals it depends on the brand.) Testing also reveals you can control how much l-theanine you coax from tea. Letting tea steep longer raises the l-theanine levels, whereas adding a lot of dairy milk reduces it.[11] Green tea does have other health benefits, such as reducing your risk of cardiovascular disease and cancer, so if you can enjoy green tea, you'll get the biggest bang for your buck.[12]

If you love iced coffee but iced tea has no appeal, try taking an l-theanine supplement with your coffee right before you need to focus. Sixty to one hundred milligrams of l-theanine is the dose researchers use to improve speed and attention. (More l-theanine, however, isn't necessarily better. Taking two hundred mg or more of l-theanine in one sitting could make you sleepy, the exact opposite of what you want.)

It's not like turning on a light switch, though, so don't expect the effects to be instantaneous. The caffeine and l-theanine take a little time to get into your bloodstream. Take your tea (or caffeinated drink plus l-theanine supplement), and you should see your focus improve within 25 to 30 minutes, with the benefits lasting at least another 60 to 75 minutes. By then, you should hopefully be deep enough into your task that you won't need additional help.

2. Grab Your Headphones

Let's say it's after 3:00 p.m. and you know that if you have even a sip of caffeine at this time of day, you'll be staring at the ceiling late tonight, wide awake, cursing that green tea and Therese Huston along with it. There must be another way to focus without forgoing sleep. Or perhaps you're thinking that you can't wait 30 minutes—you really do need to get going *now* (or better yet, an hour ago).

The next strategy comes straight from the archives of the weird, but after reading the research, I've tried it many times and it consistently zips up my focus immediately. You're going to listen to something called binaural beats. If you're an etymology buff, you're already guessing this word refers to two sounds or two ears (binary + aural), and in both cases, you'd be correct.

Here's the idea: You put on headphones or earbuds and listen to two different tones played simultaneously. One ear hears a tone at one frequency while your other ear listens to a sound that's at a slightly lower frequency. It might seem like it should be annoying to hear two different tones played simultaneously, but surprisingly, it's not. Your brain automatically subtracts one tone from the other and what you actually hear is a third tone that's the difference between the two. So if one ear is listening to a tone at 450 hertz (Hz) and the other is listening to a tone at 410 Hz, the resulting tone is 40 Hz (450–410 = 40).[13]

The experience? It's like listening to a steady humming noise, like a machine in the background. It's not particularly catchy or interesting, which is a good thing, because you don't want it to distract you.

Instead of grabbing your attention, the sound should allow you to direct your focus where you want it. Scientists are finding that listening to binaural beats has a variety of cognitive benefits, and one of the most consistent is that it helps people create a spotlight of attention. When these beats are playing, people find it easier to move their attentional spotlight where they want it while distractions fade into the background. So you should be less tempted by your Amazon cart and Slack channels, and your mind shouldn't wander as much.[14]

Finding Your Optimal Beat

As with many other findings in this book, scientists are clear that binaural beats help focus attention, but they are less clear on precisely *why* this works. The current hypothesis is that these beats do something called "entraining" the brain. Entraining basically means that you're synchronizing brain waves so that neurons fire at the same frequency. Neurons in your cerebral cortex can fire at different frequencies or tempos, as slow as 0.5 Hz (or half a cycle per second) and as fast as about 90 or 100 Hz (or 90 to 100 cycles per second).[15] Frequency corresponds, roughly, to the intensity of cognitive activity, so if your brain is firing around 1 Hz, you're in deep sleep.[16] Those super-slow oscillations are called delta waves. If your brain is firing at 35 Hz or more, in contrast, those fast oscillations are called gamma waves, and you're doing heavy mental lifting. At around 35–40 Hz, you're problem-solving, you're concentrating, you're doing what you came here to do. So when you want to focus, you ideally want your brain to be firing at 35–40 Hz.

Binaural beats help achieve that optimal firing frequency. Neuroscientists use electroencephalography tests, or EEGs, to measure what binaural beats do to the brain, and they often find that the brain will quickly fall in step with the frequencies of the tones being presented. (If you've ever seen a movie in which a character is wearing what looks like a Medusa swimming cap with dozens of wires snaking out of it, that person was hooked up to an EEG machine.) So if you are presented with two tones and the difference between those tones is 35–40 Hz, you're basically tuning your brain into its optimal focus frequency. It might seem as though it would be simpler to just present a single tone at 35 Hz—then you wouldn't need headphones and you could even create high-focus environments, perhaps by playing that 35 Hz tone in the background of a meeting to get everyone to focus up. But one tone doesn't work. In fact, one study found that a single tone, even at 40 Hz, made people more susceptible to distraction, not less.[17] Something about the subtraction that the brain does with the two tones is crucial to the efficacy.

You can find binaural beats by either downloading an app or doing a quick YouTube search. You'll find that almost all apps (and many YouTube recordings) offer background music and nature sounds while the binaural beats hum in the background. You might enjoy the music, but if you're trying this for the first time, go with pure tones and no music because pure tones have been more rigorously tested. Remember, you're trying to focus on your most important task, not the music. (You can always try it with music later to see if that's your preference.)

One thing to bear in mind with this strategy is that the numbers matter. Here are two guidelines:

- To focus your attention, you want to tap into what's called the gamma frequency.[18] To do that, look for binaural beats at 40 Hz. Studies that use binaural beats at less than 40 Hz are hit-or-miss—sometimes they improve concentration, sometimes they don't. It's not that higher is always better, but 40 Hz appears to be the sweet spot.

- Find the duration that's right for you to put you in your focused zone. Set a timer so that you listen to the binaural beats for 15–25 minutes, then turn them off. Trust me—longer isn't necessarily better.

Sometimes when I start an hourlong binaural beats recording, I find that I am successfully focused and lost in the work, and then suddenly I feel the slightest bit queasy. When I pause the recording, I often discover that it's been on for 25 to 30 minutes. But by then, I'm fully focused on my work, so I keep the beats turned off. Incidentally, the queasiness stops as soon as the sound does.

If the first recording you try doesn't work for you, try another. I found that listening to one particular YouTube recording at 40 Hz was perfect for me, but another grated on my nerves. I can't explain the difference, but I will say that you should try a few before giving up. And play with the volume, too. You want to be able to hear it, of course, but personally, I play it more quietly than I would play music.

DOES AGE MATTER?

There's good news and there's bad news when it comes to aging and one's ability to focus. The good news is that our minds wander less as we age, so we are less likely to zing from one random thought to the next.[19] When you're working on something, your own thoughts are probably less of a distraction than they used to be, which means you have the potential for incredible focus.

The bad news? Although your internal world becomes less distracting as you age, the outside world becomes more distracting than ever. Your ability to focus peaks sometime in your early thirties to early forties and then begins to slowly decline, but you'll probably notice the biggest change in your sixties.[20] Between age sixty and seventy, a sharp change occurs for many. It's as though distractions in the outside world hijack your attention all the time.

See if this sounds familiar. You pick up your phone to check the weather, but another app grabs your attention and a few minutes later when you set your phone down, you suddenly remember that you wanted to check the weather. You walk into your bedroom to retrieve something, your attention is grabbed by something else, and only when you leave your bedroom do you remember what you meant to do. Or you find yourself preferring quiet, subdued restaurants now, even though you once enjoyed bustling, noisy ones. You may chalk these things up to other aspects of aging—your memory or your hearing aren't what they used to be—but they can also reflect your level of distraction. Those other apps on your phone tempt and distract you. Objects in your bedroom grab your attention. The loud volume in the restaurant, be it music or other diners, distracts you. All of this makes it harder to focus and do whatever you're trying to do.

What strategies are most helpful for improving your focus if you're over sixty? You can try any of the methods in this chapter—they shouldn't hurt—but if I were you, I wouldn't buy a five-year supply of l-theanine. L-theanine plus caffeine reduces mind wandering, and if you're older, that's less likely to be your biggest challenge when it comes to focus.

The biggest focusing challenge for most healthy older adults is that as we age, it becomes harder to filter out distractions and inhibit irrelevant information, which means several things are competing for your attention at once.[21] When you were younger, it was probably easier to zero in on one thing and ignore those competing distractions, but when you're older, ignoring isn't as easy. You can try binaural beats if you're working on a task and want to reduce the intrusive tug of the external world. Having headphones on will reduce outside noise, of course, but the focusing beats should help as well. One unpublished study with adults over the age of sixty-five found that binaural beats at a lower frequency (they used 11 Hz) was highly effective at improving focus.[22] So if 40 Hz doesn't work for you, try a lower frequency, ideally 10–12 Hz. I haven't seen this finding replicated (or even published), but it doesn't hurt to find the frequency that gets you in a distraction-free zone.

There's also a brain-training app you can try. Please note that 99% of brain-training programs aren't worth the money. These training programs make you highly skilled at their specialized games, so you feel like you're improving, but the skills rarely generalize to improvements in everyday life. At the time I write this, however, there appears to be one exception. It's called BrainHQ, a brain-training program that was first launched by Dr. Michael Merzenich, a leader in brain plasticity research. I was highly skeptical at first, but it's been shown to help older adults improve their ability to focus their attention in over a dozen independent research studies.[23] You can try some of BrainHQ's free exercises online before you buy. What's nice about BrainHQ is that it should improve your ability to focus even when you can't wear headphones, which, let's face it, is during most of life.

---

3. Stare, But Not into Space

There's one more trick to try. The research was conducted in Taiwan with elementary school children but was based on an ancient Chinese tradition used with adults.[24]

It's called "fixation focus training," and it's almost laughably simple. Set a timer for 1 to 3 minutes. Pick a small object in front of you that isn't moving and that you're willing to focus on. In the original research study, the researchers had children stare at a spot on the wall, but when I do it, I simply open up a new window in my internet browser (opening a new window ensures I have a blank white screen and no ads or images to catch my attention) and then I stare at the crosshatch that appears next to the new tab. Start the timer and keep your attention narrowly focused on that crosshatch or whatever object you've chosen until your timer goes off. It's okay to blink. Just keep zooming your vision in on that spot. When the timer goes off, you should find that you have renewed focus and that it's easier to ignore distractions when you start working on that project you really care about.

In the original study, the research team had the children move around while they focused. They balanced on one leg at a time. They squatted, stretched, and slapped their bellies with alternating hands. The researchers added these activities because there is ample evidence that physical activity, particularly activity that requires coordination, can improve attention and focus.[25] You can try some of those activities as well, just remember to keep your eyes focused on one spot. And perhaps close your door if you might be distracted by activity in the hallway. The whole point, remember, is to improve your focus. (They also had the kids focus for much longer, approximately 20 minutes, but who has time for that outside elementary school?)

Why does fixation focus training work? The neuroscience behind this simple strategy isn't clear yet, though the research of Dr. Emily Balcetis at New York University indicates that narrowing your visual focus helps you attain your goals more often.[26] Perhaps it helps because when you narrow your focus, you're getting practice refocusing and ignoring everything else, and that practice trains you to refocus and ignore nonvisual temptations too, such as the ping of your phone or the conversation in the hallway.

Staring at a crosshatch for 2 minutes? It might be the easiest hack in this book, and if it gives you laser like focus, it's staring for the win!

TRY THIS
Your "Get Focused" Toolkit

▶ **Drink tea.** Caffeine plus l-theanine should give you a focused calm, help you resist distractions, and help you work faster. If you can't stand tea, take a 60–100 milligram l-theanine supplement with your cup of joe.

▶ **Grab your headphones.** Find a 40 Hz binaural beats recording on YouTube that you like, one that just has tones without any music, and listen through a set of headphones for 15–25 minutes.

▶ **Stare, but not into space.** Set a timer for 2 minutes and keep your gaze focused on a single nonmoving object for the full time. I usually open up a new browser window and stare at the crosshatch on the tab. It's okay to blink, but keep your focus steady. When the timer goes off, return to your project with renewed focus.

▶ **Does age matter?** If you're 65 or older, binaural beats at 10–12 Hz might work better for you than 40 Hz. Try both to see what helps you more. And although most brain-training apps aren't worth the money, BrainHQ is the rare find that's been rigorously tested and can help improve your attention.

CHAPTER 2

Get Creative

Your manager has said she'd like to meet tomorrow to generate some innovative solutions for handling customer complaints, and she's asked you to come ready with several ideas. You've set aside 2 hours this afternoon to brainstorm on your own, but 50 minutes have already ticked by and you're still staring at a blank page. You start to sweat. You've tried ChatGPT, but none of its suggestions were terribly original. (Since when is "starting a customer service team" innovative?)

How do you tap into your most creative side? Should you wear all-black tomorrow, a trademark of creative types?

Wear whatever you want (wearing black doesn't seem to help) but if you truly want to get inspired and imaginative, here are five things you can do.

You May Not Be Salvador Dalí, but You Do Create

First, let's define what it means to be creative. Psychologists define creativity as generating something that is both novel and useful.[1] A novel idea that isn't useful, such as proposing that your company have the cast of *Frozen* sing "Let It Go" at the start of every customer complaint call, won't win you points.

Even if you don't think of yourself as creative, chances are you use this skill more often than you realize. You need to be creative at work when you're asked to cut spending, coach a colleague, draft a document, or find a loophole. You're likewise creative at home when you're trying to figure out what to do with leftover rotisserie chicken, where to go on a cherished three-day weekend, or how to help your tween earn some spending money.

What Works

Thankfully, psychologists have discovered a number of ways to increase your creativity.

1. Buy Yourself a Bouquet

One approach is to put flowers on your desk. Several studies reveal that simply having a plant or some flowers nearby helps people generate more creative ideas.[2] Of course, if you only have an hour left to brainstorm, you don't have time to run to a florist, but if you need to be creative on a regular basis, invest in some greenery. I bought a succulent for my desk after learning this tip, and I keep it just to the right of my computer screen so I can see it when I'm working. It survives my neglect beautifully.

Similarly, in the warmer months, I try to write out on my back deck. Instead of enjoying a single plant on my desk, I'm surrounded by green trees blowing in the breeze and flowers in our garden. I'm a little more easily distracted out there than I am at my desk, so it's not great for focused editing, but I do love to brainstorm out there.

2. Go for a Walk

Another thing you can try is a walk. Many historical figures famous for their creativity, from Steve Jobs to Virginia Woolf, walked to get their creative juices flowing.

Psychologists are finding that science supports the stroll. Research reveals that people who walk more generate a greater number of creative ideas than people who walk less.[3] You can be what psychologists call a "chronic walker," which sounds like a terrible affliction but simply means that you walk daily, or you can be an "acute walker," which means you mostly go for walks right before you need to sit down to create. Chronic walkers tend to be more creative than acute walkers, so try to follow in the footsteps (pun definitely intended) of Jobs and Woolf, and make walking a regular part of your day.

But since you need to be creative *now*, push away from your desk and start walking. A quick jaunt will help more than sitting still. A recent study finds that you need to stride at least 100 steps to see an initial uptick in creativity, but ideally you should walk about 1,200 feet (about a fifth of a mile) to noticeably improve your brainstorming.[4]

There are two more ways to make walking work for you. First, go for a "random" walk, one in which you try a new path and explore a bit, rather than taking a set route you've done many times before. Researchers find that people who walk a path they know are less creative than people who walk a spontaneous, unscripted path.[5] Although walking has been the most rigorously tested activity, any kind of unrestricted movement generally seems to stimulate creativity.[6] If you can't walk for some reason, close your door and move your upper body, head, and eyes. Or simply try some TikTok dance moves from your desk chair.

Second, after walking, brainstorm *before* you get distracted. In most research studies, individuals showed their creativity boost within 5 minutes of completing their walk, so you want to start brainstorming either while you're walking or as soon as you stop. I've become a fan of capturing ideas on my phone either while I'm strolling or right as I finish. It's easy to assume that a great idea is here to stay, but like a Fourth of July sparkler, it often lights up, sizzles for a few moments, and then snuffs out. You don't want to bump into someone in the hallway after your walk, chat a few minutes, then discover you can't reconstruct your brilliant idea.

If you want a quick dip into one of the many research studies on how walking benefits creativity, do a quick internet search for a TED Talk on the topic. As Stanford researcher Marily Oppezzo observes in her popular talk, you aren't going to necessarily create the Sistine Chapel from start to finish after a single walk, but you will stimulate the initial brainstorming process.[7]

And if you're a runner, that helps creativity too. In chapters 5 and 6, we'll examine how more intense aerobic exercise like running improves your ability to think, but for now, know that in most studies of chronic walkers, creativity researchers actually measured the total *number* of steps, not whether those steps were made by walking or running. So, if a run is more your style, slip on your sneakers, belt out a quick quarter mile or more, then get back to brainstorming.

3. Watch a Funny Video

One of the reasons you might be struggling with creativity is that you're feeling anxious and you're trying too hard. If you've been working for almost an hour and you've got nothing worth sharing yet, the pressure is on. And anxiety typically hinders creativity.[8]

A little distraction and a good mood, on the other hand, commonly increase creativity. So set aside your project for a few minutes and watch some funny videos on YouTube or Instagram. Researchers find that a funny video both improves a person's mood and stimulates their creativity. The key here, though, is to not watch too many videos and procrastinate for too long, which will simply increase your anxiety all over again. The sweet spot, at least for increasing creativity, seems to be about 7–8 minutes of viewing.[9] So set a timer, laugh out loud, and then try brainstorming again.

Why would watching a funny video lead you to more creative ideas? One possibility is that you're creating a pleasant incubation period for yourself. When a person is feeling pressured and anxious, they often pursue the first idea that comes to mind, crappy or not. They're thinking, "I need to do something and here's an idea, so let's go with that." But if you force yourself to step away from the idea and incubate it, your mind has a chance to work on the problem in the background. When we subconsciously percolate on a problem, we often restructure it, finding new ways to approach it, which can lead to insight.[10] This valuable incubation period might be part of why taking a brief walk works as well.

DOES AGE MATTER?

If you're the only person over 60 in a team brainstorming meeting, you might be tempted to sit quietly, believing that the most creative, fresh ideas are going to come from the young people in the room. Not necessarily.

One thing that can actually serve older adults well, at least when it comes to creativity, is that, as we saw in chapter 1, aging makes it harder to inhibit distractions. Think of inhibition as your brain's filter, working to block irrelevant information that's competing for your attention. But distractions, annoying though they may be, have an upside. They can spur creativity. Researchers find that older adults often perform better on creativity tasks than their younger peers because distractions give them new ideas.[11] People in their twenties are successfully shutting out unwelcome intrusions, but because older adults find that hard to do, those intrusions seep into the creative work they're doing and new ideas often take hold.

How might this play out in real life? Imagine that you sit down to brainstorm ideas to improve customer service, and you see an email about an upcoming HR workshop titled "Psychological Safety Improves Team Productivity." You're not exactly sure what psychological safety is, so you close your email and return to your brainstorming task. You're still thinking about this notion of "psychological safety," though, and you wonder, "What if customers feel unsafe or judged when they call in with a problem they can't solve? Could we help them feel more safe and less judged?" Research suggests your younger colleagues, who are more successful at shutting out distractions, would be less likely to make this connection.

So the next time you're in a team brainstorming meeting and you need to be creative, don't count yourself out because of your age. Your distractions might be your hidden superpower.

The only problem with these first three approaches to increasing your creativity is that neuroscientists haven't yet figured out *why* they work. We don't know what the brain does differently on a random walk or near a vase of tulips. But researchers have identified two more promising ways to get your creative juices trickling if not all-out flowing, and they have a clearer sense of why these strategies work.

4. Eat Chicken or Edamame: Dopamine to the Rescue

One of the most well-documented ways to increase your creativity is to increase your dopamine. Dopamine is a neurotransmitter that's produced in your brain and allows the neurons in your brain to communicate with one another and the rest of your body.

Perhaps you've heard that dopamine is the "feel good" chemical, and it does often make you feel good, but it's really the "let's do this" chemical. We'll learn more about dopamine in chapter 3 on how to get motivated and in chapter 11 on how to feel less pain, but for now, it's sufficient to know that you produce dopamine, and it helps you spring into action.

Dopamine also plays a role in creative thinking. Do you have any friends that you think of as highly creative? There's a decent chance that, compared to the average person, those friends have more gray matter in the areas of their brains that heavily process dopamine.[12] Just as car aficionados usually have a bigger garage where they can tinker, creative people often have bigger brain areas devoted to dopamine where they can play around with ideas.

Researchers are finding that if your dopamine levels are low and you bring them back up to normal, you'll get a boost in creativity. The simplest, most convenient way to boost your dopamine is to consume an amino acid called l-tyrosine.[13] It's the protein your body needs to make more dopamine, and in general, foods that are high in protein are also high in l-tyrosine. The table below lists the estimated l-tyrosine levels in foods that should be easy to grab for lunch or a snack.

FOOD	SERVING SIZE	AMOUNT OF L-TYROSINE (mg)
Lean beef	3.5 ounces	1700
Chicken breast	3.5 ounces	1300
Tuna	3.5 ounces	1240
Turkey breast	3.5 ounces	1170
Edamame	4 ounces	650
Tofu	3.5 ounces	550
Cottage cheese	4 ounces (½ cup)	550
Cheddar cheese	1 ounce	550
Greek yogurt	6 ounces (¾ cup)	450
Chickpeas	4 ounces (½ cup)	375
Almonds	¼ cup	260

You could also take an l-tyrosine supplement, which is how researchers increase dopamine in the lab, and 2.0 grams (or 2,000 mg) of l-tyrosine appears to be an effective dose when it comes to creative work. (Lower amounts of l-tyrosine might also work for boosting creativity—they just haven't been tested—so don't feel you need to eat 2 cups of almonds.) Most supplements take effect quickly, within an hour, so you could take a supplement, work on a mundane task like email for 40 minutes, and then return to your creative task, ready to generate fresh ideas.

Dialing In the Dopamine Sweet Spot

Note that I said, "if your dopamine levels are *low*." You don't want your dopamine levels to be too high. The relationship between dopamine and creativity is a bit like Goldilocks and the porridge—you want your dopamine levels to be just right, which in the case of the brain is not too low and not too high. Too little dopamine and your creativity suffers, but too much dopamine, and your creativity also drops, as illustrated in the diagram below.[14]

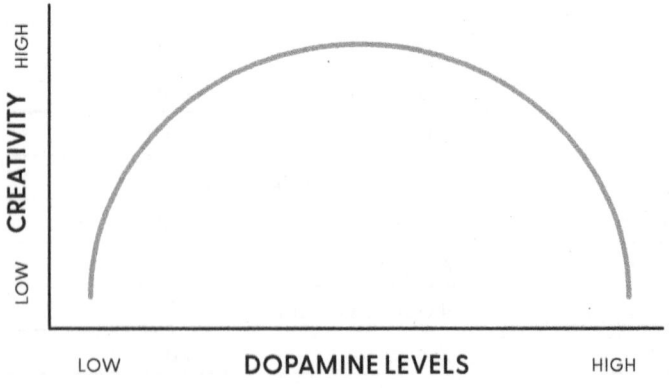

How would you know if your dopamine levels are low? Well, how stressed are you? High stress levels deplete neurotransmitters such as dopamine. If you're highly stressed, perhaps about tomorrow's meeting with your manager, then your dopamine levels have probably taken a hit.[15] (If you've ever wondered why it's easier to be creative when you're relaxed on a Saturday night than when you're stressed on Monday morning, now you know part of the story.)

You can also consider your sleep the past few nights. If you're sleep-deprived, you probably have unusually low levels of dopamine and find it harder to be creative. We've all had days when we haven't had enough sleep and we find ourselves sitting silently in brainstorming meetings, drawing a blank while others seem to pop with ideas. On mornings when you haven't slept enough, a few turkey roll-ups would be the snack to pack.

But on days when you're well rested or relatively relaxed, consuming a lot of l-tyrosine could make your levels of dopamine too high, which could tank your creativity. You definitely don't want to max out your dopamine levels when you connect with your manager tomorrow in that brainstorming meeting. In addition to hurting creativity, high dopamine levels are linked to poor impulse control, which means you might be tempted to interrupt your boss every few minutes or blurt out the first thing that comes to mind.[16] Not a good look.

Incidentally, you might be thinking, "Should I really eat turkey if I'm sleep-deprived? I thought turkey was supposed to make you sleepy." Yes, there is tryptophan in turkey, which can make you sleepy, but it's in small amounts in most turkey. There's typically more l-tyrosine than tryptophan in cooked turkey, and that dopamine boost from the l-tyrosine should make you more alert. If you're sleepy after eating a lot of turkey on Christmas Day, chalk it up to a huge meal, any alcohol you drank, and the high levels of fat and carbohydrates in all those side dishes and gravy. (And in my family, it doesn't help that my aunt cranks up the heat to 25 degrees Celcius.)

5. Remember a TV Show: Ted Lasso to the Rescue

This last strategy is going to seem like it can't possibly work, but hear me out. Researchers at Stanford, Harvard, and the University of Auckland discovered it, and they've found that it improves creativity on multiple tasks.[17] It's another way to utilize videos to harness your creativity, but instead of watching one, you're going to remember one.

Get out a pen and paper or open a new blank document on your computer. Now think of a movie or TV show that you've watched in the past month. It doesn't have to be a show you liked, but choose one with characters and a storyline, rather than, say, a nature documentary. And it can be completely unrelated to your work. You can try something funny, like *Ted Lasso*, or something intense, like *The Whale*.

Now set a timer for 5–6 minutes and start writing down as many details as you can about the show. First, describe the setting. Capture several key settings where the story takes place and be specific about what they looked like. Were the walls in Ted Lasso's office covered in wood paneling or just painted cinderblock? Next, describe some of the characters. Go beyond listing the names and describe what you saw—what was the lead actress wearing in that episode, did she have a distinctive hairstyle, was she sporting black high heels or red high-tops?

If you have time, recall the plot. Pick a memorable scene and describe what was happening to whom and why.

When the timer goes off, take a deep breath. Now try brainstorming again. It might take a few minutes, but you will likely discover that new ideas finally start breaking through.

Not Just Any Peanut Butter Memory

This might seem like we're just trying to put you in the same relaxed state you were in when you watched the show. Relaxation wouldn't hurt, but when you recall all those details, you're actually doing something very specific and intentional with your brain that makes it easier to generate creative ideas.

You're engaging what psychologists call your "episodic memory." You have many types of memories, but two of the most fundamental are semantic and episodic.[18] Semantic memories are like your own personal dictionary. If I ask you, "What is peanut butter?" you'd use semantic memory to explain how it's a thick spread made of ground-up peanuts, kids love it with jelly, and, if you're editorializing, chunky is better than smooth, *thank you very much.*

Episodic memory, on the other hand, is more like a detailed journal entry, in which the memory has personal details attached. If I asked, "When was the last time you had peanut butter?" you might say, "I actually had some this morning, standing in the kitchen, when I smeared a spoonful on an English muffin. I was going to make myself avocado toast, but the avocado on the counter had gone bad . . ." Episodic memories, like this one, have a specific time, place, and narrative arc.

When you're trying to recall a specific episodic memory, like your improvised breakfast or the show you binge-watched over the weekend, you fire up two key regions of your brain. First, you activate your hippocampus.[19] The hippocampus lies deep in the center of your brain (you actually have two, one on each side). It looks like a stretched out C lying on its back, and it sits just above your ear, about 1½ inches inside your brain. The hippocampus plays many roles, but it's especially important in memory. When you retrieve a specific personal memory, the hippocampus fires up. (We'll hear more about the hippocampus in chapter 6, when we dive deeper into memory.)

If the hippocampus were to fire all by itself, your mind would probably wander, like a pinball zigzagging between different episodic memories. You might recall your breakfast one moment, then a text you got from your mum while you ate, and then, for no apparent reason, realize that you're out of cat food. And that's why, when you're trying to bring up a very specific memory, you need a second brain region to help you out.

The second region that fires up when you go searching for a specific episodic memory is known as the executive control network. The executive control network actually isn't a single area; it's called a network because it involves several brain areas firing together. Just as an entire orchestra can play more complex pieces than a single instrument can, when an entire brain network fires, you can do more complex tasks than when a single region is working by itself. The executive control network is activated when you're trying to achieve a goal and it's important to suppress other thoughts or memories that could distract you. It helps you focus on the specific details of the TV show you just watched and prevents you from spinning off into a dozen trains of thought.

Priming the Pump

Okay, so now we know which brain areas engage when you're trying to recall a specific memory: the hippocampus and the executive control network. Lovely. But how does that help you think creatively?

Neuroscientists are finding that these two regions that light up when you're recalling an episodic memory, such as a TV show, also light up when you're trying to think creatively. Specifically, when you're trying to generate a variety of new ideas, you activate your executive control network as well as your hippocampus. These two areas that help you reconstruct the past also help you imagine the future.

You might be thinking, *Hmm, these two activities seem completely unrelated.* In one case, you're remembering something you saw or did or experienced—something that actually happened in your past; in the other, you're trying to generate new and useful ideas—something imaginary that *could* happen in your future.

But keep in mind that new ideas are often just bits and pieces of old ones that have been combined or applied in a novel way. Imagine that you're trying to think of something creative to do with your son or nephew on a rainy Sunday. You suddenly remember a zoo exhibit where he crawled through an underwater glass tunnel while crocodiles swam overhead and he was entranced. Light bulb moment. You suggest that he make big pictures of crocodile underbellies and tape them to the walls and ceiling in his bedroom to create his own underwater crocodile cave. He runs to get paper and crayons and you pat yourself on the back for your inventiveness. You used memories of a past, unrelated experience—a day at the zoo—to create a new possibility. As Mark Twain is often credited with saying, "There is no such thing as a new idea. It is impossible. We simply take a lot of old ideas and put them in a mental kaleidoscope. We give them a turn and they make new and curious combinations."[20]

Likewise, when you're thinking of innovative ways to handle customer complaints, you could generate a creative idea by recalling a seemingly unrelated memory that offers a fresh insight. You know most people in the customer service department are already trying hard, but what would motivate them

TRY THIS
Your "Get Creative" Toolkit

▶ **Buy a bouquet or plant for your desk.**

▶ **Go for a walk.** A brief walk, ideally 1,200 feet or more (about ⅕ of a mile), can immediately boost your creativity. Daily walking helps even more, as do meandering walks.

▶ **Watch a funny video.** Watch funny videos for 7–8 minutes to elevate your mood, reduce your anxiety, and give your mind a chance to incubate the problem. Just don't distract yourself and procrastinate for too long or your creativity will drop again.

▶ **Eat turkey or edamame.** If your dopamine levels are low, potentially because you're stressed or sleep-deprived, boost them by consuming l-tyrosine, either through a even more? Then you remember, out of the blue, that the energy company for your home has a clever motivating strategy. Your energy company has "flex events," in which they designate a date and time and encourage you to use less energy for 3 to 4 hours. For that period, you get ranked against 99 other similar households in your area on your energy consumption. The first time you participated, you placed 31 out of 100. In the next flex event, you were motivated to get into the top 20 homes (meaning you used the *least* amount of energy). You redoubled your

supplement or by eating a food high in l-tyrosine, such as turkey, chicken, and edamame.

▶ **Remember a TV show.** Set a timer for 5–6 minutes and write down as many details as you can recall from a TV show or movie you watched recently. Describe the plot, the characters, or a scene in vivid detail, then immediately return to the task that requires your creativity. Because you're priming the brain areas responsible for creativity, you're likely to generate some new ideas.

▶ **Does age matter?** You might find it harder than you like to ignore distractions, but those distractions can be a source of inspiration and help you generate more creative ideas than your younger peers.

efforts and reached your goal, plus you discovered strategies for saving energy that you now use regularly. The whole experience was highly motivating.

What if you did something like that with your customer service team? There could be a flex event competition once a week or month that lasts a few hours, and employees would be told how they ranked during that event based on the average customer rating on customer service calls during that window. People would be motivated both to try new approaches (it's only for a few hours, so what's the harm?) and to see their ranking improve. And like you with your energy consumption, they might discover some strategies for handling calls they could use regularly.

What's brilliant here is that our creative prompt, "Recall a show you watched recently in detail," is much more effective than simply asking people to "think creatively" without any other warm-up. Researchers find that the movie or TV show prompt works remarkably well because nearly everyone has a show they've watched recently, and it effectively primes the two key brain areas needed for creativity.[21] Then, when you give these two brain areas a new goal—generate solutions for customer complaints or a rainy Sunday idea for your son—it's already properly warmed up and you're more likely to remember an experience that could spur fresh ideas. You're essentially priming the pump, making it easier to jump into creative mode.

CHAPTER 3

Get Motivated

You need to write a report that you've been avoiding. You keep thinking about how challenging this task will be and the reasons not to do it. It involves extensive data analysis, which doesn't come easily to you. You also aren't an expert on the subject matter—someone else usually writes this report, but they're out on leave and you drew the short straw. As if that wasn't bad enough, you doubt whether anyone besides your manager will actually read it (and you're not even sure that she will).

When you go to your manager and ask, "Please, can't you give this to someone else?" she just shrugs and says, "It doesn't need to be brilliant. Just do your best."

Your motivation is at an all-time low. Is there some way to stoke your interest? What if the difficult parts didn't seem, well, so difficult?

We all have times when our motivation sags. You might have to do a task that doesn't align with your goals or interests, that feels too hard or complex, or that's just plain boring. This chapter is about bolstering motivation when we face a task we dread.

WHAT DOESN'T WORK

Don't say to yourself, "This task doesn't really matter. I can do a crappy job and no one will know or care." That framing might reduce your performance anxiety, but it won't motivate you to get the job done. It might work for a task that you could complete in 5 minutes—say, writing a quick email—but you need a different framing for a big project. Otherwise, you will feel unmotivated every time you go to work on it.

Part of the problem here is that your goal, however you slice it, is a lousy one. It's either to (a) get out of the responsibility altogether or (b) "just do your best," but don't bother with brilliance, as your manager put it. Psychologists find this "do your best" approach might be popular among parents and managers but is, in practice, highly unmotivating because it sets a low bar and is too vague.[1] Research indicates you'll be much more motivated to act if you can identify a specific, more challenging goal that you're working toward.[2]

What Works

1. Find the Bigger, More Challenging Goal

To clarify, when I suggest that you find a more challenging goal, I'm not suggesting that you, say, write a longer report using only statistical analyses that you don't know how to perform. The goal isn't to make this task harder than it actually is, but instead to see how doing this task that doesn't matter to you moves you toward achieving some other specific, challenging long-term goal that *does* matter to you.

To see how this works, let's delve into an example. How might this report help you reach a goal you *do* value? Maybe there's a dataset you have always wanted to understand and that you'll have to dig into as you write the report. Or maybe you are trying to become more efficient (and less paralyzed) when it comes to data analysis. Or maybe you could use this report as evidence in a negotiation with your manager. Perhaps you want a promotion and if you can write a report that summarizes the department's current and future needs effectively, you can argue that you're ready to handle a bigger team and more strategic responsibility. In any case, it's important that you find a specific, relatively challenging goal that does matter to you, one that this otherwise unpleasant task will help you move toward. (More on the benefits of setting difficult goals in chapter 4.)

2. Harness Dopamine: The "Let's Do This" Chemical

To understand the neuroscience of motivation, we need to delve into dopamine because dopamine creates a feeling of wanting and the desire to act. As we learned in chapter 2, dopamine is a neurotransmitter. More specifically, dopamine is a neuro-modulator, which means it changes the activity of other brain cells. Dopamine can be excitatory and increase the activity of other neurons, or it can be inhibitory and decrease the activity of other neurons. Essentially, dopamine presses the accelerator or the brakes in certain parts of the brain and nervous system. We're particularly interested in the excitatory role of dopamine as it can motivate you to act.

Let's take a quick look at how dopamine works. There are several dopamine or dopaminergic pathways in the brain, and they regulate everything from anticipating something you enjoy to controlling involuntary body movements. (If you know someone who has Parkinson's and struggles with hand tremors, you've seen the result of a dopamine deficit. Those hand tremors are partially explained by the fact that dopamine is no longer inhibiting those involuntary movements.)

When we talk about motivation, we're interested in the role dopamine plays in a brain area called the ventral stria-tum. The ventral striatum is in the very center of your brain, slightly above your ears, but deep in the middle, slightly for-ward of the hippocampus, an area we learned about in chapter 2. The ventral striatum is important for identifying whether something you've experienced or done was rewarding so that you know whether you want to seek it out again. The first time you experience the pleasure of an activity, say eating your first spoonful of Ben and Jerry's ice cream, part of your brain releases dopamine and your ventral striatum reacts to it. And when your ventral striatum gets a burst of dopamine, it feels good. That dopamine release is both motivating—you probably want to dive in immediately for more—and it helps you learn something valuable, namely that Ben and Jerry's is worth buying again.

In the future, if you liked the experience enough the first time, all it will take is someone suggesting that you buy a pint of Chunky Monkey and, bam, you'll get a little dopamine release. When you're anticipating getting or doing something that has been rewarding in the past, dopamine increases in the ventral striatum. Dopamine will also be released when you're anticipating a song you like to listen to, a task at work you like to perform, or a person you like to see. It puts you in a state of "let's do this," especially for things you've enjoyed before. If your kid's mood improves as soon as you pull into the parking lot of their favorite store, that's dopamine doing its job.

Your Dopamine Dilemma

Returning to your work dilemma, you can see how you've really got two challenges. First, since you haven't liked data analysis or report writing in the past, there's probably not a lot of dopamine flooding your ventral striatum right now. Lower dopamine levels are part of why you're feeling unmotivated. The second problem is that to feel better and raise your dopamine levels, you might be tempted to do something more immediately rewarding, like checking your Instagram feed or your stock portfolio or, *Ooh, look, I just got a DM.* You've learned that these activities can be very rewarding some of the time, so you get a hit of dopamine simply thinking about them. But then you start checking Instagram, and that pulls you down a rabbit hole and 15 minutes later, you still haven't started your report.

As Dr. Anna Lembke, the author of *Dopamine Nation*, puts it, "dopamine is about wanting, not about having."[3] If you anticipate watching an Instagram video, your dopamine goes up, so you check your feed and watch a few, but you might find yourself always wanting to watch another, because dopamine makes you want more. The trick is to raise your dopamine levels and make you want the *right* thing—in this case, to work on the report—not the wrong thing (one more cute otter video).

Let's get dopamine working for you, not against you.

Tapping into Your Inner Piglet: How Dopamine Changes Your Focus

Since dopamine makes you want something highly desirable and familiar, it's easy to see how it causes you to gravitate toward something you know that you already like, such as eating a favorite ice cream, playing Exploding Kittens, or lying in your backyard hammock. You've relished those things in the past, you've learned they're rewarding. But how is dopamine going to make you want to do something difficult that you've never done, like writing that report?

An international team led by researchers at Brown University discovered part of the answer.[4] These researchers first measured how much dopamine each person had available in their brain (specifically in the ventral striatum). Some people had low levels of baseline dopamine, others had high levels.

The researchers then gave participants one of three pills: a placebo or one of two drugs that increased the circulating level of dopamine in the brain. Once the drugs had taken effect, they showed participants descriptions of several different memory tests along with the advantages and disadvantages of each test, including each test's level of difficulty. The participants got to choose which test they wanted to take. Some were easy to perform while others were difficult, but the harder the task, the more money participants could earn. So there was a small but real incentive—more money—to pick the harder task.

What's interesting is that participants with higher levels of dopamine, either naturally or because of the drugs, chose to do the harder task more often than people with lower levels of dopamine. Dopamine drove harder choices. What's even more interesting is that the researchers monitored precisely where participants were looking before they chose. They found that those with higher levels of dopamine spent more time looking at the advantages of the harder test (the money they could earn) before they made their choice, whereas people with lower levels of dopamine spent more time looking at the disadvantages (how hard the task would be). And not surprisingly, if you're only focused on how hard a task will be, you're not motivated to do it.

In other words, dopamine acts as a dial, adjusting your attention to the advantages and disadvantages of a task, even one you've never done before. If your dopamine levels are low, you can't help but ruminate on the hurdles. You're like gloomy Eeyore from *Winnie-the-Pooh,* thinking, "Is it *really* worth all that effort?" But if your dopamine levels are high, you're more like hopeful Piglet, believing, "Why, yes, it really *is* worth the effort." With the right levels, those challenges you'd normally avoid suddenly become more appealing.

Listen to Music That Moves You. Since your dopamine levels are probably low for this dreaded task, let's look at a few ways to raise them. We learned one strategy for increasing dopamine in chapter 2—ingest some l-tyrosine. It's an amino acid that helps your body produce dopamine, so as you try to stoke your interest in diving into writing the report, you could eat foods that are high in l-tyrosine, such as beef, pork, chicken, turkey, and tofu, or you could take an l-tyrosine supplement.

But there's another simple strategy you can try without rummaging through your refrigerator or popping any pills: Listen to music that gives you chills. A fascinating study by a research team in Quebec found that when adults were either anticipating hearing music they loved or listening to that music, they had a dramatic increase in the amount of dopamine in their ventral striatum, a key area of the brain when it comes to feeling rewarded and motivated.[5] I'm sure you knew that music could move you, but you may not have realized that it can also motivate you.

In their research, the neuroscientists didn't just use iconic pieces of classical music that everyone would recognize or the top 40 hits on Spotify. Each participant hand-picked a favorite piece of music to listen to, one they claimed "gave them the chills." It was *their* chosen piece of music only, not someone else's favorite, that elicited the dopamine release.

It turns out the "chills" are important. When you have that reaction to a piece of music, your body goes through a cascade of changes, from faster breathing to a drop in body temperature. Some researchers speculate that the pleasurable chills you feel when you listen to powerful music reflect the dopamine release you've just experienced.[6]

This gives new meaning to the idea that music *moves* us. Your brain releases dopamine when you hear a song on the radio you absolutely love, and that dopamine motivates you to literally move toward a hard goal. (Personally, I now make a mental note whenever a song or music video gives me chills. It then becomes part of my "dopamine toolkit.")

Take the Plunge (or Better Yet, Take off Your Parka). One of the trendiest ways to increase dopamine is to take a cold shower. This strategy is often called "deliberate cold exposure." Immersing yourself in cold water has been shown to increase levels of dopamine. It can be as simple as turning off the hot water in your morning shower and keeping your body in the cold water stream for as long as you can stand it. Or you can fill your bathtub with cold water and plunge right in.

Uncomfortable? Mercy, yes, it's supposed to be. According to a classic study that everyone cites, the young men (mean age of twenty-two) in the study sat in 57-degree Fahrenheit water (14 degrees Celsius).[7] That's roughly the temperature of tap water, so it's not frigid. But to submerge your body in it? You probably take your normal hot shower at 100–105 degrees Fahrenheit (37–41 degrees Celsius), so 57 degrees is *much* colder. It's roughly the temperature of the Pacific Ocean off the coast of Oregon in September. Yeah. It's the kind of water you'd normally wear a wet suit for or, if you're in swim trunks, you'd probably wade in then wade immediately back out, shouting, "It's too cold!"

How much will you increase your dopamine if you can last 15 seconds in a bracingly cold shower? It's hard to know. In the classic study, those robust young men sat on lawn chairs in a swimming pool of cold water up to their necks for an uncomfortable 60 minutes, and blood tests indicated that their dopamine increased by 250%. At the time I write this, I haven't seen any studies that measured dopamine changes for adults in their forties who stood in a 60-degree shower for 15 seconds, which is, let's face it, more realistic for most of us.[8] And how long do the effects last? Again, it's hard to know. In the classic study I've just described, dopamine levels peaked at 1 hour after the young men toweled off and were still high 2 hours later, but again, these men had been in a cold plunge for a full hour.

I have tried cold showers in the morning, and they are incredibly invigorating. I towel off and am pumped to do any-thing—let me at it. But most days, it's just not realistic for me to employ this strategy exactly when I most need it. If it's 3:30 in the afternoon, say, and I'm having trouble getting motivated for my last task of the day, I'm not going to step away from my desk, undress, take a cold shower, get dressed again, and return to my desk. Or what if you're working in your cubicle downtown but you discover you need a motivation boost? I don't know about you, but I've never had a shower in the same building as my office.

So I've created a different strategy. I often do deliberate cold exposure by taking a cold walk. I aim to get out there when it's 1–7 degrees Celcius, and, where I live, I can do this five months out of the year. I wear light clothing (usually a long-sleeve T-shirt and thin leggings), tie a coat around my waist (if something happens and I'm stuck out there a long time, I want a way to warm up), and don't wear a hat or gloves. And then I take a 10-to-12-minute walk. About 3 minutes in, I am cold. Six minutes in, I am *very* cold and by 8 minutes, I'm really uncomfortable, which is, in this case, a good sign. I come home incredibly energized and motivated for whatever is next.

Best of all, I can do this from home, from my office in the city, or at a winter conference when I need to get energized for a long talk I'm about to sit through. Admittedly, I haven't seen any data on how much a cold walk will increase your dopamine, but I also haven't seen any data on the 15-second cold shower, and the cold walk is, at least for me, much more realistic. Plus, as we learned in chapter 2, a walk can enhance your creativity, so you get two benefits for the price of one.

Drink a Big Cup of Joe. Another way to give your body a dopamine boost is, you guessed it, caffeine. Caffeine doesn't actually increase your brain's production of dopamine. Rather, researchers have used PET (positron emission tomography) technology to show that caffeine increases the number of neurons in your ventral striatum that are receptive and responsive to dopamine.[9] (When I say "increases the number of neurons," I don't mean that caffeine causes you to sprout new neurons; rather, caffeine is a short-term fix. Existing neurons in the ventral striatum that previously weren't responding to dopamine will now be excited by dopamine.)

Because caffeine makes some neurons more receptive to dopamine, more dopamine can bind to these receptors, and the net effect is that one of the primary reward centers of your brain will get a strong signal to pursue something, making you a little more interested in whatever goal is top of mind. You'll feel more alert from the caffeine, yes, but also, the apathy you were feeling beforehand will be replaced by a *wanting* to do something, a desire to act. Then you need to direct that desire toward writing that pesky report, not toward, say, a vigorous pursuit of something on eBay.

How much caffeine do you need and when should you consume it? It can take about an hour for caffeine to take full effect, so you probably won't see an immediate uptick in your motivation, but if you grab a cup of coffee on your way to lunch, you should be raring to go soon after you get back to your desk. Unfortunately, the research team didn't test different doses. (PET studies are expensive and involve injecting people with radioactive substances, so as a scientist, you don't ask folks to come back to the lab five times just to satisfy your curiosity.) Researchers only tested 300 mg caffeine in their study, which is roughly the amount in the following drinks:[10]

BEVERAGE	CAFFEINE
16 oz. Starbucks Grande coffee	310 mg
16 oz. Bang Energy drink	300 mg
24 oz. Dunkin' Donuts Iced coffee	297 mg
16 oz. 7-Eleven brewed coffee	280 mg
1½ 5-Hour Energy shots	200 mg each

Personally, this table frightens me. I'm extremely sensitive to caffeine, so ingesting 300 mg in one sitting would render me unable to sit any longer. (*And why is everyone talking so slowly? And I need to pee.* The list of side effects goes on.) I love my morning caffeine, but about 150 mg over 3 or 4 hours is all I can handle. If you're like me and don't drink much caffeine, lower doses will probably still increase your dopamine receptivity and motivation, so try a dosage that won't make you jittery.

And for those of you at the other end of the spectrum, for my friends who can drink three Red Bulls and go back to the fridge for more, try to keep it to 400 mg a day or less. Dietary guidelines in the U.S., Canada, and Europe specify that 400 mg is the healthy upper limit.[11]

3. Do a Self-Affirmation, but Not the Stuart Smalley Kind

You might be wondering about "motivation apps." There are plenty out there and many provide you with a daily affirmation or inspirational quote. Do self-affirmations actually boost motivation?

Research reveals one kind of self-affirmation does help, but not the generic "You can do this" or "You are strong and capable" messages so common on these apps. These steady drips of encouragement may lift your mood and may even be your favorite way to start your day, but unfortunately, the research doesn't indicate that they make you any more likely to work on that damn report.

So what is the type of self-affirmation that will motivate you to tackle work you're dreading? It was originally developed at Stanford University and University of California, Santa Barbara. Here's how it works: First, ask yourself what you value most in life. No, it doesn't have to be remotely related to your job. If you're having trouble identifying what you value most, look at the list below and pick one value that you'd say, yes, that matters more to me than most other things on the list. Most items probably matter to you somewhat, so if you're having trouble picking just one, ask yourself, "When I think about a day I'm really looking forward to, what's the focus of that day?" and use your first thought as a guide. There's no judgment here.

- Friends / Family
- Religion / Spirituality
- Leisure / Hobbies
- Health / Fitness
- Travel / Adventure
- Creativity
- Humor
- Nature
- Independence
- Career

If your top value isn't on this list (and it's by no means exhaustive), please feel free to write in your own.

Now, think about a time in the future when you might spend time or resources investing in this value, what that might look like, and why it matters to you. So, if you picked friends and family, you might think about how you'll be having friends over for dinner next week. Don't just think about your friends and move on. You need to reflect on this for several minutes. Grab a notebook or open a blank document on a device and spend 4–5 minutes jotting down specific ways you'll soon spend time with your friends and why they're so important to you.

When your 5 minutes are up, close the notebook. Now start on whatever work you were dreading. There's a good chance you'll feel more motivated.

I know. It seems crazy that thinking about how you'll invest time in your hobbies or health could help you with something completely unrelated, but neuroscientists found that this kind of values affirmation activates the ventral striatum and brain areas associated with drive and reward.[12] They didn't measure dopamine, so we don't know if dopamine increased, but we do know the key brain areas responsible for feeling more motivated sprang into action. (They also found that a week after this self-affirmation activity, the people in the study were exercising more, even though no one had ranked health, fitness, or exercise as their top value. Thinking about how much they cared about nature or their family was enough to motivate them to do something entirely unrelated that they knew they *should* be doing, namely taking care of their health.)

Why does this work? By anticipating future rewards, you're activating your ventral striatum, which gives you momentum. This kind of value-affirming activity also puts you in touch with a broader view of yourself and reminds you why you're here. When you remind yourself of your core values, it becomes easier to move past your current hurdles that might be threatening your sense of self.[13] You remember that you're competent and bring value to the world, which helps you realize, for example, that a difficult report isn't the end or the boss of you.

One key note: The researchers found that thinking about a *past* experience when you embraced your values didn't activate the ventral striatum, perhaps because a key role of the ventral striatum is to anticipate future rewarding experiences. So, if you picked friendships as something you value, don't think about the lunch you had with your best friend yesterday. You can be grateful for that time together and enjoy those warm fuzzies another time but it won't motivate you now.

Instead, imagine a future lunch date, even with a less-precious friend, in exquisite detail and why it matters to you. Just don't text that friend when you're done. Start on that report! You can thank your ventral striatum (and your friend) later.

TRY THIS
Your "Get Motivated" Toolkit

▶ **Find the bigger goal.** Identify a bigger goal that does motivate you, a goal that this dreaded task can help you achieve.

▶ **Harness dopamine.** Increasing dopamine, the "let's do this" chemical, can strengthen your drive and steer your attention to the advantages of doing a hard or undesirable task. Try one of these three approaches for an immediate dopamine boost.

> **Listen to music that moves you.** Take a few minutes to listen to music that gives you chills.

> **Take the plunge or take off your parka.** Try deliberate cold exposure by standing in a cold

shower for as long as you can or by going for a walk, minimally dressed, in the cold.

> **Drink a big cup of joe.** Consume 300 mg of caffeine via a beverage or supplement.

▶ **Do the right kind of self-affirmation.** Take 5 minutes to do a values affirmation. Identify a core value that deeply matters to you and write about a concrete time in the future when you'll be investing your time or resources in that particular value.

DO YOUR BEST WORK

CHAPTER 4

Accomplish More

You'd like to move the needle on your dreams and accomplish more. A lot more. You've searched online and tried strategies that the "most productive" people do—like taking more breaks and tackling your hardest tasks before lunch. Those tricks work for your urgent tasks—the high-pressure ones with deadlines and other people counting on you—but what about those side projects that don't have immediate deadlines and if you don't make progress on them, no one will care but you?

You keep putting those personally important projects on your to-do list, starring and underlining them, but you never make any real headway. The urgent tasks always win.

Most people I know have a big dream or goal they struggle to make time for. Maybe it's a project at work that their manager has green-lit as long as they complete their other work. Perhaps it's a side business they'd love to launch. Or a YouTube channel.

There are only so many hours in a day, though, and you're probably already squeezing in a lot. The struggle is real. And research indicates that desperately wanting to achieve a goal doesn't necessarily help you reach it.[1] After all, abandoning a New Year's resolution is as traditional as making one in the first place.

There are plenty of productivity gurus out there, but neuroscience has some unconventional advice. When it comes to achieving your big goals, there is one thing you shouldn't do and four things you should.

WHAT DOESN'T WORK

Oprah clearly knows how to get stuff done (could she be more amazing?), but you need to ignore a popular piece of advice on her website if you want to realize a big dream: Don't make a vision board.

Self-help websites and life coaches often tout vision boards as a way to "manifest" your dreams. Here's the typical advice: Look for inspiring images and phrases that capture what you're trying to achieve or where you're trying to go, arrange those pictures artfully on a posterboard, and hang them where you'll have a visible reminder of your ideal outcome—a picture of where you're headed, and how fulfilled you'll feel once you get there.

On the surface, it makes a lot of sense. You can't achieve what you can't see, right?

Yes and no. Having concrete goals is important, and we'll say more about that in a moment. But focusing too much on the ideal outcome you want and how incredible you'll feel once you get there can backfire and actually demotivate you. (So much for *The Secret*.)

Researchers have learned that if you imagine an ideal, positive future for yourself, you lose your drive.[2] Participants in one study who pictured an ideal outcome felt less energized and enthusiastic about their goal than those who imagined a neutral outcome. When people fantasized about a rosy future, their blood pressure also immediately dropped, and blood pressure is a reliable indicator of how energized and enthusiastic you feel. (We normally think of low blood pressure as a good thing, and that's true for overall health, but when you're picturing a future event that's going to take some effort, a short-term increase in blood pressure is more desirable. It shows that your body is amping up and preparing for the hard work.) What's particularly troubling about this result is that it was the participants who said that they wanted the goal the most, those who had the greatest need for achievement, who had the biggest drop in blood pressure when they imagined their perfect future.

Reduced drive, not surprisingly, also affects your immediate output. When people entertained positive fantasies about the outcome they wanted and pictured it in vivid detail, they accomplished less in the following week than those who simply wrote down a few thoughts about the week ahead.[3]

Vision boards could also have another unpleasant side effect down the road: You're putting yourself at risk for depression. Psychologists have found that positive fantasies about the future can make you feel good and reduce depressive feelings in the present, but if you're fantasizing about impressive accomplishments, that can actually lead to *more* depressive symptoms later.[4] People were more depressed if they had imagined a perfect future in which they'd attained all their goals than if they had envisioned a negative or balanced future, and the more positive their fantasies about their achievements, the more depressed they were six months later. One reason that people with positive fantasies were more depressed was they spent significantly fewer hours working on one of their goals than people with neutral or negative views of their future, and as a result, they made less progress. Time moved on, but they remained in place.

The takeaway is clear: If it's something you need to work toward, don't picture the work as complete.

Why would visualizing your rosy future in all its glorious details backfire? One theory is that by positively fantasizing about your future, you mentally experience those juicy rewards *now*, rather than needing to wait until you've done all the work. The neuroscience supports this view. Researchers have found that brain areas typically associated with high levels of reward, certain parts of the amygdala, are activated when you mentally simulate a positive future.[5] It makes you feel amazing in the moment, but since you're *already* feeling what you want to feel, why work so hard to actually get there?

What Works

1. Picture the Process

It seems as though the lesson from "What Doesn't Work" is that you should avoid visualization if you want to be a high achiever. That's not quite true. Done the right way, visualization can help you achieve your goals. Here's the trick: Picture the *process* of moving toward the goal you want, not the goal itself. Researchers find that envisioning the process of working toward an outcome leads people to spend more hours working and, as a result, increases their chances of achieving that outcome.[6]

This can all feel a bit abstract, so let's step through an example. If your goal is to, say, write a book, picture yourself coming home from work, changing into your comfiest clothes, making yourself a cup of tea, sitting down at your computer, and writing. See yourself setting your phone where you won't be tempted to check it. You picture your entire routine, all the way up to closing your laptop once you've written a page or two.

So, the good news is that visualization works. The bad news is that effective visualization is a bit boring, at least compared to the fun of making a colorful vision board. Picturing the process, at least for most work tasks, isn't sexy, but it *is* effective.

The "Fine, I'll Do the Right Thing" Brain Area

Brain science can help us understand why envisioning the process works. At first glance, it seems as though it works because you create an action plan, and planning is half the battle. That could be the entire story, but neuroimaging actually tells a much more interesting tale . . .

A team of researchers at Harvard and Cornell used functional magnetic resonance imaging (fMRI) to examine what happens in the brain when people imagine several steps they would need to take to achieve a goal. They then compared it to what happens when people picture the plausible and enjoyable events that might occur after they've achieved that goal.[7] Let's say the goal was to eat healthier. The people who were asked to picture the steps might imagine buying a bunch of organic vegetables and chicken breasts, cooking a delicious recipe with those ingredients, and eating fruit for dessert while everyone else eats cake. We'll call this first group the planners. The people who were asked to picture the plausible outcomes might imagine themselves fitting into a favorite pair of jeans, taking more selfies, and seeing lower numbers on the scale when they weigh in. We'll call this second group the winners.

The research team discovered that different brain areas fired up for the two groups. As we mentioned earlier, the winners— the people who imagined the glorious outcome—activated areas associated with feeling rewarded and fabulous. And why not? It feels incredible to look good in clothes you thought you would never fit into again.

But what about the planners? The people who imagined the process of working toward their goals activated many more brain areas—five times as many, in fact. It takes more work (and more brain) to plan out all those steps. I won't go into all the brain regions that lit up for the planners, but I'll highlight one that's telling.

The area that showed the greatest increase in activation for the planners compared to the winners was the dorsolateral prefrontal cortex on the right side of the brain, toward the front. (Yes, dorsolateral prefrontal cortex is quite a mouthful, so most neuroscientists just call it the right dlPFC, although I'm not sure that's any better.) In general, the dorsolateral prefrontal cortex is important for holding multiple things in memory simultaneously, and if you're planning a healthy meal, you can imagine how you might be thinking about many components, from the grocery list to the cooking time to who is going to complain that you're serving kale.

But the dorsolateral prefrontal cortex on the right side? That's special. It's like the finger-wagging area of the brain. That right side is especially active when you're deciding you'll finally do what you *should* do, not what you *want* to do.[8] The right dorsolateral prefrontal cortex lights up when you see more tempting options and yes, you feel conflicted, but you still choose the hard road.

And deciding that you're going to choose the hard road—the thing that doesn't bring you as much pleasure now but will bring you greater fulfillment later—is crucial to achieving challenging goals. You need to say, "I'll have a bowl of berries instead of cake," or "I'll work on that book chapter for 30 minutes instead of watching Netflix." You need that finger-wagging brain area to be firing on all cylinders. It's no wonder that imagining all the specific steps they would take and activating that particular brain area made it more likely that people would work toward and, even more importantly, reach their goals. So picture the process and get your brain to help you, not hinder you.

2. Avoid Alphabet Soup: Set Goals Differently

Visualizing your process, boring though it may be, is the one goal attainment strategy that has clear neuroscience revealing why it works. So if we were being brain science sticklers, we would stop there. End of chapter. But psychology research points to several *other* strategies that increase one's chances of reaching a goal, and they're well documented, so I'll share them even though, at the time I write this, there isn't much neuroscience explaining how they work.

Does the way you set a goal improve your chances of reaching it?

Some of you are probably nodding vigorously, thinking, *Of course it does, Therese. You need the right goal-setting framework to set your performance metrics.* Maybe you're a fan of OKRs (objectives and key results). Or maybe you're in the KPI (key performance indicators) camp. Depending on the organization you work for, you may have OKRs, KPIs, or SMART goals (Specific, Measurable, Achievable, Realistic, and Time-Bound). Yes, acronyms abound in the goal-setting world.

Although many organizations have adopted one or more of these goal-setting frameworks (it's sometimes said that Google grew from 60 employees to 135,000 employees because it aggressively adopted OKRs back in 1999), there is little scientific data indicating that any of these frameworks leads to *consistent* behavior change. Are people more likely to achieve their goals once their team sets their OKRs? Do individuals accomplish more after they have defined KPIs than they did before? When surveyed, managers indicate they like OKRs, and think they lead to more effective teams,[9] but opinion surveys often have a lot of baked-in bias. After all, if you've spent several days every quarter creating and evaluating new

OKRs, you'll likely think they're effective simply because you've invested so many hours and you don't see yourself as a time-waster.

Setting acronyms aside, there *is* scientific research revealing what kinds of goals are most likely to lead people to change their behavior. A meta-analysis (i.e., a study that averages the results of other scientific studies) of over 140 research papers reveals that the most effective goals tick all three of the following boxes:

- They are difficult but achievable.
- They are set publicly, not privately.
- They are set as a group goal.[10]

One finding was particularly counterintuitive—difficult goals were more desirable. In this meta-analysis, the research team found that people who set harder goals were more likely to achieve their goals than those who set easier goals. Obviously, an impossible goal such as playing in the NBA when you're fifty years old and only five feet tall wasn't going anywhere, but as long as the goal was achievable and people were given enough time, they were more likely to accomplish those challenging goals. So aim for something that's a stretch but still within reach.

One reason people are more likely to achieve difficult goals is that easier goals are often unmotivating and require minimal steps.[11] The goal "drink a glass of water every morning" is easy to do, but pretty boring. Fill glass, drink, done. In contrast, "run my first 5K" is not only more interesting, it's harder to do and requires training and commitment. There are many steps to running your first 5K (literally and figuratively), and as we've seen, imagining and planning out the steps you need to take makes you more likely to achieve your goal. Plus, you feel inspired and motivated each time you run a little farther and get closer to your bigger goal.

As for setting "public" goals, you can simply share your goal with a partner or a close friend. With social media, just about any goal can be set publicly. Or you can go old-school and write your goals on a whiteboard in your office where others can see them. A friend of mine does this, and she even puts a running tally next to her top goal to indicate how many days in a row she's spent at least 15 minutes working on that goal (I see this as the knowledge-worker's equivalent of the "Days since last accident" signs). As soon as she misses a day, painfully, she must reset her counter to zero. Her friends love to peek in, see the tally, and celebrate or commiserate. I love this system so much I sometimes ask her to send me a picture of today's tally.

Setting a group goal may be the least achievable item on this list. Most of the time, our personal goals, such as launching a YouTube channel or cleaning up the piles of paper in our office, are not group goals, but we can enjoy some of the benefits of a group goal by using a coaching app. There are apps such as GoalsWon and Coach.Me in which you set your goals and work with a virtual coach to help you reach them. You get a brainstorming and accountability partner, and at the time I write this, they cost as little as £2 a day. (In an independent rating of 344 behavior change apps, social scientists gave Coach.Me a relatively high score for incorporating a wide variety of techniques that lead to successful behavior change.[12] GoalsWon offers similar features, but it was launched after the ratings were published.)

So take a minute to think about your goal. If it's too easy or if you set it privately, or if you're the only one working toward it, now you know what needs to change.

3. But What If I Want a Nap? Create If/Then Guardrails

There are two more strategies that every serious goal-seeker needs. First, research indicates that you should do if/then planning. You need to think through different scenarios and be able to say, "If X happens, then I will do Y."

Let's say that your goal is to write a book. It's been a burning desire for more than a decade, but you haven't made much progress beyond scrolling through your favorite authors' websites for advice. That's reading, but sadly, it's not writing.

If/then planning, or "implementation planning" as psychologists like to call it, means you think through several different scenarios and how you'll respond in each of them. If/then planning is particularly effective if you think through two possibilities: (1) The opportunities you'll have to work on your goal, and (2) the temptations that may pull you away from it.[13] Your if/then statements for writing your book might be:

- If I take the bus to work and I find a seat (opportunity), I'll write for at least 15 minutes.

- If my spouse takes the kids on a weekend afternoon (opportunity), I will use that time to write.

- If I am working on my book and I start to feel sleepy (temptation), I will run up and down the stairs a few times to wake up and then get back to my writing.

Some if/then statements focus on a specific time and place, whereas others focus on how you feel or how you'll interrupt yourself when you start getting off task. You don't need 30 of these—you won't remember them all anyhow. Start small, perhaps with three well-chosen if/then statements that will help you make better use of your time and give you a plan for those moments when you are likely to get derailed. Capture them in a place where you can see them frequently. You can try a daily reminder app such as Remember the Milk, where you can easily set when and where you'll receive your if/then reminders.

I like to think of if/then statements as guardrails—safety mechanisms to catch me if I get off track. My favorite if/then guardrail is this: If I've been at my desk for more than 20 minutes but haven't started working yet, then I'll put on headphones and listen to binaural beats (for more on these beats, see chapter 1 on getting focused). I keep a pair of Air Pods on my desk for this very purpose; it locks in my concentration every time.

Research indicates that if/then planning is highly effective. Compared to individuals who don't use it, people who employ this hack regularly are much more likely to reach a wide variety of goals, from, say, doing self-exams for testicular cancer to, yes, completing a major writing project like a book.[14]

4. Monitor Your Progress the Right Way

This last strategy is simple and incredibly effective. Yet, oddly, research indicates it's a tactic people often avoid. If you want to reach your goal, try monitoring your progress. You want to regularly note what you've accomplished and how that compares to your goal. It can motivate you to see how far you've come, help you identify when you need to try a different strategy, and ensure you keep your goal top-of-mind.

Two things make progress monitoring successful, say the lab coats.[15] The first builds on something you just learned—making your progress (not just your goal) public. Research indicates that sharing your progress with at least one other person helps, so ask your partner or spouse if you can do weekly check-ins. Social media can also help here, as can a coach. My friend's office whiteboard tally might be extra effective because she's got a public record of not just her goal but also how many days in a row she's worked on it.

You can also improve your chances of reaching your goal by physically recording your progress. You might shrug and say, "But I mentally review my progress during my morning shower." That might make you feel better but it doesn't help you get more work done. You need to write your progress down. That might mean using an app or having a tally in a notebook. Personally, I love Excel spreadsheets, and I create one for nearly every project that is more than a week long, but *you do you*.

There is one tweak to this strategy that's not obvious. When you first start working on a goal, focus on how much you have already done, not on how much more there is to do. If, for example, you're preparing a presentation, note that you went from 0 slides to 4 today. If you want to run your first half marathon, log when you go from running 10 minutes to running 20 minutes. This is emotion regulation at its finest, and research indicates that focusing on the 1% you've completed will motivate you more than focusing on the 99% remaining.[16] (If that sentence depresses you, I've made my point.) Ride the pride in what you've done.

But once you cross the halfway point on your project, shift your focus to how much is left. Let's say you want to apply for a master's program to further your career, and you're more than halfway done with the application. Focus on the fact that there are just two essays you still need to write, not that you've spent 30+ hours requesting recommendations and transcripts and filling out application forms. Most of us tend to fixate on one thing or the other for too long—either how much we've done or how much more there is left. Researchers find that when individuals shift their focus midway, they make more early progress, they maintain their momentum, and they are more likely to reach their goals.

So if you want to be at your most productive, don't adopt one strategy for checking off your progress and expect it to work for the entire arc of a project. Be sharp by changing strategies as you get closer to the finish line.

TRY THIS
Your "Accomplish More" Toolkit

▶ **Picture the process.** Rather than creating a visualization of the rosy outcome of completing your goal, which is what most vision boards do, picture the more mundane process of working on your goal, step by step.

▶ **Set goals differently.** Set big goals that are challenging but achievable, share them publicly, and try to make them group goals rather than individual goals.

▶ **Create if/then guardrails.** Identify opportunities that will support you in attaining your goals as well as temptations that will derail you, then create if/then statements to guide you through those moments. Start with just three if/then guardrails.

▶ **Monitor your progress the right way.** Write down your progress and share it publicly on a regular basis. At the start of a project, focus on how much you've accomplished and when you're halfway through, switch gears and focus on how much you have left.

CHAPTER 5

Think on Your Feet

Let's say you have a big presentation at 10:00 a.m. for a tough crowd. Everyone will be there. Your boss. Your boss's boss. That guy from R&D who seems to relish playing an aggressive game of "Stump the presenter."

You've prepared your slides and your talk is polished, but you're still nervous. What if someone asks a question you didn't anticipate and your mind goes blank? What if the technology doesn't work and you can't show any of your slides? What if you fidget with your hair, something you often do when you're nervous? Most importantly, how can you think on your feet, nimbly and deftly, so that people are persuaded by everything you've worked so hard to prepare?

You need strong executive function for a big presentation. Executive function is what cognitive psychologists call those higher-level mental skills that allow you to do complex tasks in unfamiliar circumstances.[1] First and foremost, you use executive function *before* your talk, as you're writing it. Let's call it Plan A. This is your game plan assuming everything goes well.

But you also use executive function *during* your talk if plan A fails you—perhaps the technology breaks down and people are starting to fidget—and you need to develop, on the spot, standing in front of all these people, an improvised Plan B. When your executive function is at its best, you're able to coordinate your mind and body to achieve your goal, no matter what the world throws at you.

Executive Function Isn't Just for Presentations

Anything with the word "executive" sounds impressive, but what is "executive function," really? For a summary of what it is and isn't, see the table on the next page.

Let's Get Flexible

Executive function actually encompasses three different types of cognitive skills.[2] First, there's working memory, which is your ability to hold several things in mind at once. It's like a little loop or scratch pad where you hold information. If you receive a 6-digit security code on your phone and you mumble it to yourself repeatedly until you type it into the proper webpage,

EXECUTIVE FUNCTION **IS NOT** ...	EXECUTIVE FUNCTION **IS** ...
... a capacity that only executives need.	... a capacity many of us need at work (and in life in general, to be honest). When you're solving a new problem and can't rely on routine, you need executive function. Whenever you're planning a new project, changing your priorities based on new information, or resolving a team conflict, you're drawing on executive function.
... the same thing as executive presence. Executive presence is more external, it reflects the gravitas and confidence you exude and the impression you make on others.	... largely internal. Executive function involves having the mental agility to think clearly and in complex ways despite being in a novel and potentially high-pressure situation.
... something that only children or those with ADHD struggle to master.	... something that anyone can struggle with.

you're using working memory (and perhaps you can see why it's called a loop). Second, there's inhibition. When you're trying to ignore anything that's not relevant to the task you're trying to perform, that's inhibition. If you receive a text on your phone from one of your kids while you're memorizing the security code and you're successfully able to ignore that text and not open it, that's inhibition. Lastly, there's cognitive flexibility. Cognitive flexibility is the ability to adapt quickly and easily to changing circumstances.

You could need all three executive function skills in your talk. If someone asks a rambling multipart question, and trust me, we all need to field those sometimes, you need working memory to keep track of what they're asking. You also need inhibition to stop yourself from fiddling with your hair or your pen; or, if a colleague happens to send a stressful email just before your presentation, you want strong inhibition so you can put that email at the back of your mind and focus on your talk.

But cognitive flexibility is probably the most important skill to have today. If you're worried about handling curveball questions, troubleshooting technology problems, or adapting to anything that's thrown your way, you need cognitive flexibility. For simplicity's sake, I'll refer to all three cognitive skills simply as "executive function," but I'm going to point you to strategies that research reveals are especially good at improving cognitive flexibility.

What Works

1. Reduce Your Stress

How can you boost your executive function on days when you need it most? One way to improve your mental agility is to reduce your stress levels.

Acute stress, which is stress you experience for a few minutes, hours, or days, does a number on executive function. It's especially damaging to cognitive flexibility.[3] Often when we're stressed and we face a problem, we can think of only one solution, and if that solution doesn't work at first, we keep trying to make it work. In that moment, we're unable to come up with another solution, but later that night, when we're relaxed, something else suddenly comes to mind. Or if you're giving a talk, you might find that a member of the audience, someone who isn't stressed, will see an obvious way forward and suggest it. Why didn't you think of that? Because you were stressed, friend, and stress makes that kind of flexible thinking much harder.

Chapter 13 has a number of strategies for handling acute stress, but one to try is a mindfulness meditation. Many studies demonstrate that mindfulness meditation, which involves purposefully paying attention to the present moment and focusing on one's breath, reduces stress.[4] The real question is, how much meditation is enough to reduce stress? Most researchers offer several intensive training sessions, or they ask participants to meditate for a month or longer, and not surprisingly, that kind of repeated practice does indeed lower stress. But some scientists have found that even a single 20-minute mindfulness meditation is enough to reduce immediate stress levels.[5] You can find 20-minute mindfulness meditations on YouTube. The soothing effects of a single meditation probably won't last long, however, so do it as close to your nerve-racking event as possible, ideally in the hour beforehand.

Of course, if you already have a favorite way to reduce stress, use it before your talk. Maybe that's listening to a few tunes on your favorite playlist, maybe that's doing today's Wordle, or maybe that's taking a long hot bath instead of your usual shower. For me, kneading dough used to be one of my favorite relaxation techniques. Believe it or not, I carved out time to make homemade yeast bread the morning I gave my dissertation defense, all with the hope of bringing my stress levels down. And it worked. I fielded questions so well that I actually got an "ooh" when I answered a particularly hard question from the most famous neuroscientist in the room.

2. The Science: Get Sweaty

You might be thinking, *Glad that worked for you, Therese, but there's no way I can bring my stress levels down. Not today.* Perhaps this presentation is peak stress for you, and no matter how much deep breathing you do (or how much dough you knead), your anxiety is high.

Another great option for improving executive function is exercise.

But not just any exercise.

Researchers are finding that moderate- to high-intensity exercise leads to immediate short-term increases in executive function.[6] First, let's define what "moderate to high intensity" means, in general, and then we'll dive into the details of what's needed to boost executive function.

Moderate-intensity exercise means that your heart is beating at 50–70% of your maximum heart rate.[7] That might mean walking three miles at a brisk pace of 4 miles per hour or playing a match of doubles tennis. According to Mayo Clinic, you'll know you've reached moderate-intensity effort if you can talk but you can't sing.[8] Your breathing is faster, but you're not completely out of breath.

High-intensity (or vigorous) exercise is, as you'd expect, harder (and let's face it, sweatier). It means your heart is beating at 70–85% of your maximum heart rate. It includes activities like jogging at least 6 miles per hour, cycling hard (14–16 miles per hour on flat terrain or, even better, on hills), or playing a singles tennis match, in which you're constantly running all over the court. You'll know you've reached high-intensity effort if you find it hard to speak more than five or six words at a time. Whether it's moderate- or high-intensity exercise, you're working hard and you'll probably break a sweat.

Researchers who study how physical exertion improves executive function have mostly focused on running and cycling, in part because it's easy to roll a treadmill or stationary bike into the laboratory, but some enterprising scientists have studied people doing more creative activities, like burpees and jumping jacks, even Zumba dancing.[9] All of these activities, even a Zumba class that has you bouncing to Beyoncé, lead to better executive function.

There's some debate as to what's best for executive function. Should you aim for moderate- or high-intensity effort? Some researchers find that you get a bigger boost in executive function if you only work out at a moderate intensity. Sounds like a clear win since that's easier, right?

But the benefit of high-intensity exercise is that the effects seem to last *longer*. After completing a hard run, for example, people found it easier to think adaptively and flexibly for *at least* 2 hours.[10] I emphasize "at least" because the researchers didn't actually wait around longer than 2 hours, and there weren't any signs that the benefits were dropping off at the 2-hour mark, so for all we know, they might have been mentally sharp for 3 or 4 hours after stepping off the treadmill.

Like Most Things in Life, Timing Is (Almost) Everything

Okay, now we know how hard you need to exercise, but if you're like me, you're wondering how long you need to be sweaty. After all, this is a workday and you don't have all morning. A good rule of thumb is that the longer you push yourself, the longer your mind will shine. In the lab, participants who did high-intensity workouts for as little as 4 minutes saw improvements in executive function 15 minutes later.[11] So if you're about to jump on a short but important Zoom call and you want to be at your brightest, run up and down the stairs in your building as fast as you can for a few minutes, if you're able. You'll dial up your ability to think quickly under pressure and impress everyone on the brief call. (And you'll probably get a mood boost as a bonus.)

If, however, your moment in the spotlight is 3 or 4 hours from now, exercise longer. It was the 50-minute high-intensity workout that made people adaptive and flexible thinkers 2+ hours after they stopped.

There is an important caveat, however. If you exercise to the point of exhaustion, you defeat the purpose. You'll be too tired going into your presentation, plus if your workout stresses you out, those stress hormones will impair your ability to think clearly. So exercise as long as you can while still enjoying it.

How, Exactly, Does Exercise Help?

Just as an executive's role at a large company is complex, executive function is also complex and several brain regions are needed to do it well. But there's one region that's doing much of the heavy lifting when you're solving new problems and that's the prefrontal cortex, an area that's appropriately named because it's at the front of the brain, just behind your forehead and above your eyes. When people are stuck on a hard problem and they rub their forehead, they're intuitively rubbing their skull around their prefrontal cortex.

Sadly, rubbing doesn't help.

But time at the gym does. Blood flow to the prefrontal cortex increases following a workout. But this spike in blood flow wears off quickly, so blood flow alone probably doesn't explain why you're still mentally sharper 2 hours after you get off your bike.

To understand why your ability to think on your feet is still strong long after your heart rate has returned to normal, we need to turn to chemical changes in the brain. Let's first talk about neurotransmitters, the chemicals that help neurons communicate. Certain neurotransmitters increase during physical exertion, and it can take 2 hours before they return to baseline. So if you jump in the pool and start swimming laps, your brain will still be swimming in these neurotransmitters hours later.

For a variety of ethical and technical reasons, it's tricky to measure real-time fluctuations in brain neurotransmitters in humans (no one wants a tiny probe lowered into their skull). But based on the science that has been done, mostly with animals and occasionally with humans, the current thinking is that two kinds of neurotransmitters, serotonin and norepinephrine, may explain why exercise helps us solve new problems.

Serotonin is usually thought of as a mood chemical, and if your brain isn't getting enough of it, you can feel depressed. Exercise increases serotonin levels in the brain, which is one reason why you might enjoy a mood boost after a good workout. But serotonin also affects your ability to think adaptively. Researchers find that the harder you work out, the higher the serotonin levels are in your blood and, correspondingly, the better your executive function becomes.[12]

Norepinephrine affects your attention and your arousal levels. When norepinephrine levels go up in your prefrontal cortex, you feel sharper, more alert, and attentive. Neuropsy-chologists find that the longer an animal exercises, the longer those high norepinephrine levels last, which suggests that the longer you run or play tennis, the longer you'll feel your mental edge afterward.[13] Researchers also find that low nor-epinephrine levels make it harder to focus, whereas moderately higher levels make it easier to focus and juggle multiple things in memory simultaneously.[14] If you're feeling mentally foggy the morning of a presentation, 20 minutes of jumping rope should give you the norepinephrine boost and mental focus you're looking for.

Scientists aren't convinced, however, that serotonin and norepinephrine are the full story, and they're looking at other chemical changes in the brain to understand why a single bout of intense exercise can make you, at least for a short while, a smarter version of yourself. One possibility is lactic acid. When you're working out harder than usual, your body breaks glucose down into lactic acid. You might think of lactic acid buildup as a bad thing—perhaps you've heard that lactic acid is the reason your muscles are sore after a hard workout, but that's actually a myth. Here's what's not a myth: Lactic acid has benefits to your brain, because it crosses the blood-brain barrier. Researchers have found a positive correlation between the amount of lactic acid in the body and improvements in executive function following high-intensity exercise.[15] That means people with more lactic acid after a hard workout showed the biggest jump in their ability to think quickly. And in general, the harder your heart is working, the more lactic acid you produce.[16] So get your heart pumping and your brain will magically start humming.

WHAT DOESN'T WORK

You might be thinking, *But, Therese, all that effort sounds so hard. Can't I just walk to work? Or maybe take my dog on a more vigorous morning walk?*

You could, but it would have to be much more vigorous than a typical walk. Most adults walk at a pace of about 3 miles per hour. (Slower than that if they're over forty, and slower still, of course, if they're stopping frequently with a sniff-happy dog.) If you still want to be sharp hours later, you need to walk much faster—about 4 miles an hour, or 1 mile every 15 minutes. That's quite a clip. Remember, it needs to be fast enough that it's hard to sing.

Low-intensity exercise, like a walk at 3 miles per hour, does boost executive function, but researchers find the mental benefits drop off quickly, after about 10 minutes. In other words, you'll generate brilliant answers to any questions you think of as you take off your coat after the walk. But in your presentation 2 or 3 hours later? By then, the mental benefits will have long worn off.

There are other pros to walking outside, of course. As we saw in chapter 2, a walk can increase creativity, so going out in the morning might help you think of a funny and creative anecdote to tell at the start of your talk. We all love a good story. And as we saw in chapter 3, if it's a brisk, cold day, going for a walk without a coat and hat should elevate your dopamine levels and make you feel charged and motivated. If you're not excited about giving your talk, that dopamine surge could help you. But a leisurely walk won't be nearly as effective at increasing executive function and helping you think on your feet as hard exercise.

Turn Your Living Room into Your Dance Studio

Like most practices in this book, I wanted to take this one for a test drive. When I started trying to exercise on presentation days, my biggest hurdle was time. I work out regularly, but on presentation days, I usually do it *after* work, not before. When I give talks or lead workshops, I spend my morning reviewing my talking points, double-checking my tech, and, let's face it, making sure I don't look like a hot mess. (Maybe men can step out of the shower looking their dressy best in under 20 minutes, but when I need to impress, I need at least an hour to shower, style my hair, and do my makeup. And a sweaty work-out would move me deeper into hot mess territory.)

Exercise subscriptions came to my rescue. No time wasted driving to the gym; just open up my laptop and go. They also allow me to exercise at home, in my jammies, regardless of the weather or daylight. (I have a huge scar on my knee from one early morning when I ambitiously went running in the dark. Lesson learned—I play it safe whenever I can now.)

There are lots of apps and options out there: Do a quick search for "best workout subscriptions" and you'll find plenty. I tried both Daily Burn and Apple Fitness and I liked them for different reasons. I realize that cost can be an issue, but there are apps that are less than £10 a month, and you probably can't find a gym that cheap.

DOES AGE MATTER?

If you're middle-aged or older, you might have noticed that you aren't as interested as you once were in trying new things. You never would have dreamed of returning to the same vacation spot in years past, but now that has appeal. Employees thirty or forty years into their career often say they'd prefer to solve familiar problems than wrestle with the unfamiliar. "The devil you know" just feels more comfortable.

And research reveals one reason this is true: Cognitive neuroscientists find that executive function tends to follow an upside-down U-shaped curve. Executive function tends to be lower for adolescents, increases in early and middle adulthood, and then drops again in older adulthood.[17] If your executive function is lower than it used to be, you'll find it harder and more frustrating to think on your feet and adapt to the unexpected, making it much more comfortable to stick with what's predictable and what you know.

But exercise can reverse the trend. When different age groups were compared, people over the age of fifty showed the greatest improvement in executive function after exercise.[18] They felt sharper and performed better. The biggest immediate boosts in executive function after moderate- or high-intensity exercise were enjoyed by people ages fifty-five to sixty-five who were ordinarily sedentary.[19] (If you're over sixty-five or if you're ordinarily active, exercising right before you need to be at your best will still give you a significant boost, but the gains in executive function won't be quite as large.)

So get out your tennis racket or your cycling shorts. It might be a pain to squeeze in a morning spin class, but it will make it so much easier to think on your feet.

---

3. Move Mindfully

If you're thinking that you don't want to get too sweaty the morning of a big presentation or perhaps you physically aren't able to engage in strenuous exercise, I have good news. There's a lower-effort technique that's consistently been shown to improve executive function: mindfulness that involves motion. We've just talked about mindfulness meditation, which tends to be done sitting still, and we'll learn more about the benefits of different kinds of meditation in chapters 6 and 8. But if you want a surefire way to improve executive function, you also want to move.

Researchers at the University of British Columbia scoured 179 different studies to identify how to improve executive function.[20] They looked at everything from making more friends at work to lifting weights at the gym. Mindfulness plus movement was the winner. Every single study they could find that combined mindfulness with movement showed improvements in executive function. Anything with a 100% track record is almost unheard of in science, but this combination seems to have it. Admittedly, only a fraction of the studies they reviewed combined mindfulness with movement, but all those studies that did found significant gains in executive function.

What mindfulness techniques involve motion? They are largely rooted in Asian traditions, such as T'ai Chi or Tae Kwon Do. T'ai Chi is a Chinese martial art known for its slow, deliberate movements. If you've ever seen a group of adults slowly and smoothly moving in synchrony in a park on a Saturday morning, that's T'ai Chi. Tae Kwon Do is a Korean martial art that moves a lot faster and involves kicking and punching, but the classes often begin with a mindfulness meditation and breathing exercises. (Although Tae Kwon Do classes are often focused on children, a quick online search can help you find one for adults.)

You might be thinking—yoga! That's a practice with both mindfulness and movement. Yes, but unfortunately yoga has mixed results when it comes to improving executive function. Sometimes it helps, sometimes it doesn't. There are many kinds of yoga (hot yoga, for example, is so different from yoga nidra that many feel the two shouldn't even share a name), and the kind that benefits executive function is called hatha yoga. Researchers find that if you already have a regular hatha yoga practice, one 25-minute yoga session should boost your ability to think quickly.[21] (For more on hatha yoga, see chapter 13.)

It's not clear why mindfulness plus movement is so effective at improving executive function. Perhaps the coordination of mind and body is key. Perhaps it's the combination of breath work and mental focus while your body is moving.

There is good news and bad news when it comes to the mindfulness plus movement solution. The good news is that if you start doing T'ai Chi or hatha yoga regularly, you won't need to wake up early on the day of your presentation to do it—your executive function should be boosted even on days you don't practice it. The bad news is that the effects aren't

TRY THIS
Your "Think on Your Feet" Toolkit

▶ **Reduce your stress.** Acute stress hurts executive function and cognitive flexibility, so if you have a favorite stress reduction technique, it's worth making the time for it. If you don't have a go-to stress relief method, try a 20-minute mindfulness meditation just before you need to be at your best.

▶ **Get sweaty.** Moderate- to high-intensity exercise will help you think adaptively and more quickly. The longer you can exercise, the longer your boost in executive function will last, as long as you don't exhaust yourself.

▶ **Move mindfully.** Combining mindfulness with movement also elevates executive function. T'ai Chi, hatha yoga, or Tae Kwon Do have all been shown to work. This isn't a quick fix, however; aim for at least six weeks of regular practice to see lasting improvements to executive function.

▶ **Does age matter?** Aerobic exercise helps adults improve their executive function at any age, but if you're 55–65 and ordinarily sedentary, you'll enjoy the biggest boost of all.

immediate. Whereas a single aerobic workout should boost your executive function immediately, trying a single T'ai Chi class on the morning of your presentation probably won't. All the studies of T'ai Chi and Tae Kwon Do were done over several weeks (the shortest was six weeks). Even the study of hatha yoga was conducted with people who had been practicing for four months or more. But if you are looking to improve your executive function long-term, a T'ai Chi class in the park might be a perfect fit.

CHAPTER 6

Learn More and Learn It Fast

It's one of those afternoons when you need to be a sponge. You're about to be introduced to a room full of important people and you'd like to remember more names than usual. Although normally your memory is decent, today you need it to be incredible.

Is there anything you can do to boost your ability to memorize new things or at least reduce your chances of forgetting? Yes. There are two things you can do to sharpen your memory before an experience in which you want to absorb as much as possible, and two things you can do after.

The Hippocampus—Your Intake Librarian

Before we dive into effective strategies for improving your memory, it helps to get a glimpse of a key brain region that allows you to form new memories. Say hello to your hippocampus. We first encountered the hippocampus back in chapter 2, when you learned that this brain area was important for remembering an episode of your favorite TV show. The hippocampus is important not just for recalling memories but also for storing them in the first place.

The hippocampus is fascinating (and complex) enough to deserve an analogy. If you think of all the memories you've accumulated in your lifetime as books in a library, then the hippocampus is like the librarian at the front desk who takes brand-new books as they arrive and tags them so that they get shelved in a place where they can easily be found later. And just as a librarian can help you find books throughout the library, the hippocampus helps you retrieve memories that are stored throughout the brain.[1] (But I do want to come clean on something: Every analogy has its limitations, and this one is far from perfect. When you shelve a book, it doesn't change much on the shelf. The pages yellow, the binding gets dusty and cracked, but the content more or less stays the same. Memories, however, are dynamic. If you stored a memory of a conversation with your mum 10 years ago, the memory you have of that conversation today will be very different from whatever you stored a decade ago. Even memories that feel incredibly vivid and real aren't perfect recordings.)

What I like about this analogy is that it helps you to see that the hippocampus isn't where you store every memory. Just as every book isn't stored in a heap on the librarian's desk, the full memories themselves aren't stored in the hippocampus. The hippocampus is a small brain area, only 1½ to 2 inches long. It's not nearly big enough for *all* the details you can recall, from the wrapper of your favorite candy bar as a kid to the feeling in your mouth of your most recent bite of chocolate. Instead, researchers at Rockefeller University recently discovered that the hippocampus is where *general* memories are stored, such as "I liked candy necklaces as a kid," but the elaborate details, such as what colors those candy necklaces came in or how smooth and cool they felt around your neck, are stockpiled elsewhere in the brain.[2]

What Works

1. Get Sweaty Again: Your Brain on Its Own Best Drug

Now that you've had a quick introduction to the hippocampus, let's turn to memory strategies. In chapter 5, we saw that aerobic exercise will help you think on your feet, making it easy to answer tough questions or resolve a team conflict. But that's not the end of the story. Aerobic exercise also boosts your memory so that you remember more and forget less.

How does aerobic exercise help? Aerobic exercise increases an important chemical called brain-derived neurotrophic factor (or BDNF) in your hippocampus.[3] BDNF is what is known as a "neurotrophin," a kind of protein that plays a crucial role in the growth, development, and survival of nerve cells in your brain. Your brain produces BDNF, so you don't need to ingest it. Think of it as Miracle-Gro for your brain. Just as Miracle-Gro helps your garden sprout healthy new plants, BDNF helps your hippocampus sprout healthy new connections. Neuroscientists have found that even sprinkling BDNF onto neurons in the lab makes them grow new branches.[4]

You want more BDNF because it promotes neuroplasticity.[5] Neuroplasticity is your brain's ability to create new connections in response to new experiences and new information. If you are meeting a bunch of new people this afternoon and you want to lock in their names, you want neuroplasticity so that your hippocampus can connect the names "David," "Devon," and "Daniel" to the right faces. If your hippocampus is humming when you meet them, the next time you see one of them, the right name will come to you more easily.

A key way to stimulate your brain to produce more BDNF is through exercise. There is a lot of research on exercise and BDNF, so much research, in fact, that I can recommend a specific protocol for ensuring high BDNF levels:

- Engage in aerobic exercise 2–3 times a week (more than that doesn't seem to help much, and less has mixed results).

- Make it moderate intensity, 50–70% of your maximum heart rate, which, as we said in chapter 5, means you can talk but can't sing.

- Exercise for at least 40 minutes each time.[6]

The good news is that if you have a long-term exercise regimen like the one above, your BDNF levels will stay high, even if you need to take a break from exercise. So if you have a few super-stressful weeks and can't squeeze in your workouts, or if you're recovering from an injury, no worries. The BDNF levels in your hippocampus should still be high for up to a month after you stop working out.

There is, however, bad news, for people looking for a quick fix. A single workout, even a hard 45-minute spin class, won't necessarily spike your BDNF levels in your hippocampus. Most studies indicate that you need consistent exercise to see the kind of BDNF increase in the brain that improves memory. (Your muscles also produce BDNF, interestingly enough, so a single workout *will* increase BDNF in your bloodstream. The problem is that BDNF doesn't cross the blood-brain barrier, so having this miracle protein in your bloodstream won't help you form new memories.)

Even if you don't have a job in which your memory is regularly put to the test, you want more BDNF. Higher levels of BDNF in the brain protect against both Alzheimer's and depression, and labs are investigating whether BDNF can be used to treat these illnesses.[7] Whether you're planning a corporate merger or simply planning your retirement, BDNF is your brain's new best friend.

For me, learning about BDNF was enough to motivate me to increase my running time from 30 minutes to 40 minutes two or three times a week. In the winter, when the weather is at its worst, I aim for two 45-minute spin classes a week. You might need to increase your workouts gradually, but remember, the aim here is progress, not perfection!

DOES AGE MATTER?

If you're over sixty, you might be thinking, "Damn right age matters!" If there's one thing that most people believe changes with age, it's memory. Many healthy older adults report that their memory has declined. They forget the names and faces of acquaintances they once knew, or they forget the point they were going to make in a conversation. These lapses can be embarrassing, and individuals often fear it will impact their work and reputation.[8]

Such memory hiccups aren't necessarily early signs of dementia or even mild cognitive impairment. They are often just the typical lapses that come with aging. (Do talk about any memory changes with a healthcare professional, however, so you can catch any potential problems early.)

If you could do just one thing to stave off normal memory loss, what should it be?

Sign up for dance lessons.

Perhaps you're just as surprised as I was to learn that. But research studies in many countries and cultures have found that dance lessons improve memory in older adults.[9] It doesn't have to be a particular kind of dance—ballroom, swing, salsa, tango, or line dancing—and it can be solo or with a partner. But going to a class does seem to be important—I haven't seen any studies that examined septuagenarians dancing at home in their kitchen.

Does it help on Day 1? Probably not. Most published studies that find improvements in memory involve attending a dance class for six months or more.[10] It's possible that a month of lessons might make a difference, but most studies look at longer stretches.

Best of all, the neuroscience reveals that you won't just learn some dance moves that you can do at your granddaughter's wedding. You'll actually *gain* more brain. Researchers have found that dancing increases brain volume through a process called neurogenesis, in which you sprout new neurons. One study found that adults ages sixty to seventy-nine who took up dancing for the first time and kept going to class for six months significantly increased brain mass in a crucial region that sends information from the hippocampus to the rest of the brain.[11] More information gets shelved in your memory library, making it less likely that you'll forget.

It might seem odd that dancing would help, but study after study finds that, at least for older adults, it's more effective than taking up other exercise routines. Part of it is consistency—people are more likely to keep going to a dance class whereas they might give up on something equally effortful that they do independently—but another part is that it combines physical, mental, and social skills. In a dance class, you have to get your heart pumping, you have to remember the steps, and you have to joke with that person you keep bumping into. The title of one research article captures it perfectly: "Dancing Combines the Essence for Successful Aging."[12]

Growing new brain tissue at age seventy? Yes, please. Pass me those tap shoes.

---

2. Meditate Again: How More Is More

You might be thinking, "These exercise and dance tips will help, Therese, with names I need to learn later, but I need to remember names *today*." Fortunately, there are indeed things you can do to rev up your memory abilities on short notice.

Before you walk into that important meeting, try a mindfulness meditation. We learned about the positive effects of mindfulness meditation for executive function in chapter 5 and we're returning to it because it's also a memory booster. Researchers find that meditating before you need to learn something new will improve your ability to take in that information now and recall it accurately later. And 8 to 10 minutes may be all you need. One research team in the UK gave participants an 8-minute guided mindfulness meditation in which they were instructed to focus on their breathing and how it felt in their body.[13] The adults who listened to the guided meditation showed significant improvements in their ability to recognize faces immediately after the meditation. In contrast, both the control group (who could think about whatever they wanted) and the group that listened to an interesting audiobook for 8 minutes showed no improvements in face recognition. Another study found that a 10-minute guided mindfulness meditation led to, on average, a 75% improvement in recall compared to the control condition.[14]

So if you want to remember more names and faces in that meeting, block off time beforehand and do a search on YouTube for "10-minute mindfulness meditation." Go someplace where it's safe to close your eyes and get mindful.

You might be like me, however, and know that typically you can't carve out 10 minutes to meditate right before an important meeting. You'll benefit from a regular meditation practice. A consistent meditation practice leads to even more robust memory and cognitive improvements than a single session and your memory will be strong even if you haven't meditated that day. Neuroscientists at New York University found that doing a 13-minute guided meditation five days a week for eight weeks enhanced both memory and attention on a variety of cognitive tests compared to a group that listened to a podcast for the same 13 minutes.[15] By week 8, the meditators were also getting better sleep and experiencing less anxiety than podcast listeners. (You might be thinking, "Maybe it was a stressful podcast," but it was RadioLab, one people liked enough that they actually listened 6 days a week, not 5.)

You're probably wondering if you need to meditate for a full 13 minutes. Thirteen is, no pun intended, a rather odd number. Every meditation app has 5- and 10-minute guided meditations but 13-minute practices are harder to find. Unfortunately, the New York University team didn't test shorter meditations, but as we've just seen, other researchers have found that an 8-to-10-minute session boosts immediate memory, so my guess is that a regular 10-minute meditation will give you lasting memory benefits as well.

What's probably more important than the length of the meditation is how many weeks you commit to doing it. The researchers didn't see across-the-board memory improvements when they tested meditators after four weeks, only after eight. So if you want to become someone who delights people because you're consistently good at remembering their names (or if you simply want to stop forgetting where you put your keys), commit to regular meditation for at least two months.

How Meditation Works Its Memory Magic

When I first learned that meditation improves memory, I figured people had sharpened memory abilities because their stress levels had come down. But that doesn't seem to be the explanation. The people who show the biggest drop in stress levels from meditation aren't the same people who show the biggest improvement in cognitive abilities.[16] The two don't seem to be related.

Instead, there appear to be four ways that mindfulness meditation boosts memory.

First, mindfulness meditation reduces mind wandering and makes it easier to ignore distractions.[17] A regular mindfulness meditation practice reduces activity in the default mode network, which we learned about in chapter 1.[18] As you might recall from that chapter, a lot of activity in the default mode network means your mind is wandering, zipping from topic to topic, but if you meditate often, this network settles down more quickly, making it easier to focus.

But meditation doesn't just quiet the brain. It also activates brain areas that support focus. Imaging research reveals that when you're meditating, a region called the basal ganglia becomes more active, which actively helps you suppress irrelevant thoughts, ignore external distractions, and focus where you want to focus.[19] And when you walk into a room full of people, you're up against dozens of distractions. There are all the voices, there's your phone vibrating in your pocket, there's the sudden realization, *Ooh, there's Elizabeth; I need to talk with her before she leaves.* So if you're able to successfully ignore all those distractions when someone says, "Hi, I'm David England from Morgan Stanley," you're already much more likely to remember he's David, not Stanley.

Meditation also increases blood flow to a key brain region next to the hippocampus that is crucial for memory, an area called the entorhinal cortex.[20] If the hippocampus is like your intake librarian, the entorhinal cortex is like a conveyor belt bringing new books to her desk. The entorhinal cortex is the part of the brain that connects the hippocampus to the sensory regions of the brain that process incoming sights, sounds, and feelings. That 8- or 10-minute mindfulness meditation probably improves your immediate ability to remember names because it activates the conveyor belt for your hippocampus, speeding up the rate at which you can take in and file away incoming information.

Lastly, a long-term meditation practice physically rewires your brain. Studies have shown that as little as eight weeks of a daily mindfulness meditation practice was enough to significantly increase the amount of gray matter in the hippocampus.[21] Commit to meditating, and you basically get a brain remodel. You're getting that neurogenesis we described in "Does Age Matter?" earlier in this chapter. You gain additional storage space, and, to push our library metaphor, a bigger desk for your intake librarian, and a brain that's more efficient at storing memories so you can find them later. Other studies of highly experienced meditators (people who have been doing it regularly for a decade or more) find that the more years you've meditated, the more gray matter you have in your hippocampus.[22]

It's rare to find any technique that increases gray matter as you age. If anything, it's normal to lose brain matter as the decades go by.[23] So the fact that with just 10 minutes a day you can keep getting smarter as you age? Sign me up. That's just over an hour a week and in exchange you get more brain. I always meant to meditate—I knew it was good for me—but it wasn't until I read about what it can do for the brain that I fully committed. I now do a 10-minute meditation several days a week. I aim for mornings, but often find myself squeezing it in at night, because the days so often get away from me. I've noticed a moderate improvement in my memory—I have fewer moments when I'm in a conversation and draw a complete blank on someone's name, for example—but, for me, the biggest benefit is that I'm much calmer and less anxious than I used to be. Research indicates that meditation is very effective at reducing anxiety, and for me, that's a monumental win because anxiety has historically been a significant challenge.[24] My memory has always been decent; my level of chill has not.

Mindfulness meditation apps make it easier than ever to have a regular practice. The 10-minute "Daily Calm" on the Calm app was my go-to at first; Tamara Levitt's voice is so soothing. More recently, I've been turning to the meditations on Apple Fitness+ because I like the simplicity of a single app for both my exercise and meditations. I also love that I can filter by theme, like focus, kindness, or gratitude. If you're looking for a free app, Insight Timer and Smiling Mind are both hugely popular and have an impressive catalog of meditations, many of them 10 to 15 minutes long.

GREAT FOR ALZHEIMER'S, MEH FOR NORMAL MEMORY

If you do a quick internet search for "memory enhancers," you'll find that there are pills aplenty. They're called nootropics or smart drugs. The makers of these drugs don't actually promise to make you like Bradley Cooper's character in the movie Limitless, who, after taking one, can access any memory he has had since birth, but the ads come close.

I wish, with all my heart, that a pill could make you a memory magician. Or at the very least double your ability to remember things, which would be so handy on those days when you have an exam or are trying to learn all those dang names. But I've combed the scientific research and I've walked away skeptical.

One supplement that is currently garnering attention is alpha-GPC. The claim is that it improves memory, and if you have mild to moderate Alzheimer's disease, then yes, the research indicates there's an excellent chance that alpha-GPC will help.[25] If someone you love is in the early stages of Alzheimer's, ask their doctor about alpha-GPC.

But those studies are taking someone who has crumbling memory and bringing them closer to normal memory. For healthy individuals looking to go from normal to exceptional memory, the jury is still out on alpha-GPC. All the scientific studies with healthy humans I've seen to date from research universities are unimpressive. One study, for instance, found that people, on average, were 10.5% faster at a working memory task when they took alpha-GPC than when they took a placebo.[26] Sounds promising, right? But read a few sentences further into that study and the researchers concluded that the results were so mixed from one person to the next that the statistics couldn't be considered significant, which means that you might spend £50 a month on a supplement that does little for your memory.

If you already take alpha-GPC and you're a fan, great, you might be one of the lucky ones. But the research available so far is clear that exercise and meditation are the real memory builders.

---

3. Test Yourself Before You Rest Yourself

There are also things you can do after an event to hold onto the information you've just learned. The simplest is to test yourself immediately. At that afternoon meeting, for example, you could take 45 seconds to step away from the crowd while you pour yourself a glass of water silently reciting, "David is from Morgan Stanley, and he's the young, meticulously groomed guy in a bow tie; Devon is from Bank of America and is the blond-haired guy with a really wide nose" while you mentally picture their faces in turn. Or you could stand at the side of the room, sipping your water while silently quizzing yourself: *Who's that? Devon. And him? David.*

Boring? Maybe. But highly effective? Yes. Numerous studies have found that testing yourself right after you learn something is one of the most effective ways to promote long-term, lasting memories.[27]

Psychologists call it the "testing effect" and it works so well, scientists have proposed dozens of theories to explain it. Right now, one popular hypothesis is that retrieving information strengthens the connection between the cue and the information you're trying to recall.[28] In this example, Devon's face would be the cue, and the information you're trying to recall is his name, so looking at him and forcing yourself to retrieve that information will make it easier to remember his name the next time. Going back to the library analogy, once you've pulled a particular book off the shelf once, it should be easier to find that book the next time. The act of recalling something from memory also seems to dampen or weaken incorrect connections.[29] If you look at Devon's face and for just a second think, "He kind of looks like my old friend Paul," you're connecting Devon's face to the wrong name.

Testing yourself on Devon's name seems to weaken that connection to the name Paul.

You might be wondering whether you're more likely to remember someone's name if you say it out loud when you meet them. It's become a popular technique, but does it work? Sort of. I haven't been able to find research on that exact strategy, but I have an educated guess, based on the research I *have* read. Simply saying someone's name as soon as you meet them ensures that you heard it right and forces you to focus up. Paying attention is the first step in creating most memories. In that way, saying a name starts the storage process—it's like setting a book on the librarian's desk—but if you stop there, it probably won't be enough. You're not necessarily storing that name in long-term memory (or, to squeeze all we can from this analogy, you're not moving it from the librarian's desk to its proper place on the shelves).

Research reveals that what helps most is what's known as "spaced retrieval."[30] The basic idea here is that you want to retrieve someone's name several times from memory, with time elapsing in between each recall. Ideally, you'll strengthen the memory the most if you're just on the cusp of forgetting someone's name and then you remember it, grabbing it just before it's gone. From that perspective, what would be more effective would be to recall someone's name 2 or 3 minutes into the conversation, saying it out loud, and then again 7, 8, or even 10 minutes later. If you're not talking with any one person for a full 7 or 8 minutes, try to refer to them by name in your next conversation. You might say, "It's funny you should mention that because Devon and I were just discussing that. Do you know Devon from Bank of America?" It's another way to test yourself.

4. Scare Yourself Smart

There's one more memory strategy you can try, and you've probably experienced it already by accident. Have you ever noticed how vivid your memories are for an emotionally upsetting event? Maybe you've seen one of your children fall from a height, and you can remember every detail as though it happened in slow motion. Or maybe you had an upsetting feedback conversation with your manager and you can remember exactly where she was sitting, the phrases she used, and even how she smelled or what blouse she was wearing, details that were irrelevant to the conversation but that your mind latched onto and filed away regardless.

Emotionally charged memories, especially ones that are imbued with strong negative emotions, are simply easier to remember.[31] When you experience something that angers, scares, or threatens you, you'll get a surge of adrenaline, also known as epinephrine. Adrenaline is one of the body's surefire ways of making certain you remember exactly what just happened. And that stamping of memory makes sense. Memories help you make predictions. If something threatening just happened, you want to remember every detail of where you were, who you were with, and what happened so that you can either avoid that situation altogether or be better prepared for that threat the next time you have to face it. (If, for example your manager smells that way again, should you be on guard? Strange though it sounds, I once had a manager who developed terrible body odor whenever she was highly stressed. It served as an effective warning system to give her space.)

Scientists find that they can induce someone to have more vivid memories of something mundane, like names and faces, by giving that person a shot of adrenaline right after they're exposed to the information they're trying to learn.[32] The timing is important. Giving someone a boost of adrenaline right *before* they try to learn something doesn't seem to work. The adrenaline has to *follow* whatever they want to learn. Adrenaline stamps in memories for events that happened before the adrenaline entered your system, not for events after. So if you had an upsetting conversation with your manager, for instance, you probably do remember incredible details from the 5 minutes leading up to her comments, such as exactly what she was doing in her office when you entered, but you probably don't remember walking back to your office afterward.

How can you use this to your advantage? You can try boosting your adrenaline right after you walk out of the meeting where you're trying to learn all of those names. In the lab they give participants a literal shot of adrenaline or they plunge a person's hand in a bucket of ice water (a surefire way to get an adrenaline surge), but that's not going to work in the office. Instead, as soon as you get back to your desk, watch a horror film clip on YouTube. (You can even have the clip cued up before you go to that meeting where you know you'll be meeting people.) Scare yourself and you'll get an adrenaline dump. But for this to work, you'll need to boost adrenaline immediately after meeting people and learning their names. If you watch that scary video after you drive home, it will be too late.

TRY THIS
Your "Learn More and Learn It Fast" Toolkit

▶ **Get sweaty.** To increase your BDNF levels, do aerobic exercise 2–3 times a week for at least 40 minutes at moderate intensity, hard enough so that you can still talk but can't sing.

▶ **Meditate mindfully.** Do a mindfulness meditation in which you focus on your breath for at least 8 minutes, preferably 10, to see an immediate memory improvement. A single session will help some, but if you can meditate 5 days a week for eight weeks, you'll see bigger boosts to both memory and attention.

▶ **Test yourself before you rest yourself.** Test yourself by trying to recall people's names (or whatever you're trying to remember) a few minutes after you've learned the information.

▶ **Scare yourself smart.** Right *after* you've learned something you want to remember, watch a scary movie clip to get an adrenaline boost that will lock in those memories.

▶ **Does age matter?** If you're over sixty, sign up for dance lessons. Stick with it for six months or more, and you'll boost brain regions associated with your hippocampus and improve your memory.

CHAPTER 7

Make Fewer Mistakes

You find yourself making the same mistake repeatedly at work, much to your frustration and embarrassment. Maybe you keep forgetting how to use an advanced feature in Zoom and fumble every time you lead a meeting. Someone kindly showed you how to use it and at the time, you thought, *Okay, got it,* but two weeks later, you find yourself turning red in the face all over again as people sit and watch (and probably judge) in silence.

Is this early Alzheimer's?

It could be, but there's a more likely culprit—attention. When you make a mistake, if you're not paying enough attention to the instructions or feedback you receive afterward or if you're busy defending yourself (internally or out loud), you'll keep making that same mistake. There are clever strategies that will help you make fewer mistakes in the future (and you'll have fewer moments of regret to boot).

Bad Math Juju

Before we dive into the neuroscience and strategies for making fewer mistakes, we need to take a step back and consider how you think about your math abilities. I know. You made your mistake in a Zoom call, not in a calculus class, but bear with me.

When you struggled with math growing up, and trust me, almost all of us struggled with math at one point or another, did you tend to think, "I'm just not good at math?" Perhaps when your parents saw you frustrated with your homework, they tried to reassure you by saying, "Don't worry—I was never good at math either," implying that you're in a family with bad math genes (or at least bad math juju).

Or, when you struggled with math, did you think, "Okay, this is harder than I expected. I need to work harder and I'm sure I'll get it"? Maybe your parents actually said something reassuring to you, like, "This is a chance to learn something new, and if you keep at it, pretty soon these hard problems will be easy for you."

The two scenes I've just described capture what psychologists call "mindsets." The first mindset is known as the fixed mindset and assumes that certain traits and abilities are relatively fixed in life. You're either good at math or you're bad at math. If you're bad at math, you may as well retire your protractor and put your energy elsewhere because, let's face the facts: You're not going to get much better. You're just not one of the lucky kids.

The second mindset is what's known as the growth mindset. A growth mindset assumes that with effort, an enthusiasm to learn, and a willingness to persevere through the hardest bits, your traits and abilities can grow over time. So you may struggle with math today, but if you take the time to learn from your mistakes, you can be good at it in the future. You may even excel.

One Mind, Many Mindsets

I bring up the math example because if you adopted a fixed mindset when you struggled with math earlier in life, you may also adopt a fixed mindset now when you struggle with something at work, math or otherwise. You've internalized the idea, perhaps from your well-intentioned parents, that there are some abilities you have and some you don't, and for the latter, don't waste your time trying. You might, for example, have a fixed mindset about public speaking skills ("I just don't have the gift of gab") or learning names ("I'm terrible with faces"). I've known people with fixed mindsets about everything from financial literacy ("bring up a spreadsheet and my mind goes blank") to feedback ("some people are good at giving feedback, but I just do more harm than good").

It helps to realize that people often have a surprising mix of fixed and growth mindsets. It's not all one or the other. Someone who has a growth mindset around learning languages and loves their daily Duolingo might have fixed mindsets in other areas. Fixed mindsets are especially common around intelligence, creativity, music, athletics, math, and technology. You, for example, might see yourself as a natural athlete (fixed mindset) and as someone who has gradually learned how to be a good leader over the years (growth mindset), but you see creativity as a gift you lack (fixed mindset). (If you think you're not creative and never will be, chapter 2 just might change your mind.)

How You View Mistakes

Let's return to the issue of mistakes and how to make fewer of them. The Stanford psychologist Carol Dweck introduced the world to the idea of fixed and growth mindsets in her bestselling book, *Mindset*. She and other researchers have discovered that your mindset changes how you experience mistakes.[1] If you have a fixed mindset, a mistake is a shameful moment. You just showed your vulnerable underbelly. You've revealed that you're bad at something, and if it's at work, you're wincing because you believe, at some level, you've just shown your managers and coworkers you will *always* be bad at that thing. (They may not think that at all, especially if they have a growth mindset about whatever mistake you just made.) If it's a skill or a task you want to be good at—or worse yet, one you *need* to be good at—you'll scurry away from the shameful moment as quickly as possible and try to avoid situations in which you need that ability in the future.

But if you have a growth mindset, a mistake is an opportunity. Failure can still be a painful experience, don't get me wrong. But as Dweck writes, when you have a growth mindset, failure "doesn't define you. It's a problem to be faced, dealt with, and learned from."[2] When you make a mistake, your next step is to lean into the problem, not away from it.

I used to do homework with a friend of mine in college who had, without a doubt, the strongest growth mindset I've ever seen. He'd be doing his physics homework, get stuck on a problem, and say, "Well, that didn't work. Oh, goodie. I love a hard problem. Now the real learning begins!" I would roll my eyes and believe he was pretending—surely no one liked to be stumped—but now that I know about mindsets, I see he had the epitome of a growth mindset.

A growth mindset doesn't mean that you believe improvement will be easy.[3] Instead, it's the belief that you have the potential for change and that you're not stuck with the ability (or lack of ability) that you have now. The only way to capitalize on that potential, however, is through effort, and it may take a little effort, or it may take a lot.

WHAT NOT TO DO

Don't get too frustrated if you make mistakes now and then. You might take comfort in what neuroscientists call the "85% rule." If you're learning something and getting it right 85% of the time but making a mistake 15% of the time, you're actually making the *optimal* number of mistakes. (I understand that it won't *feel* optimal, of course, especially if you're making those mistakes publicly.)

I say it's optimal because your learning will be accelerated: Neuroscientists have found that learning progresses the most quickly when error rates are right around 15%.[4] If you're never making mistakes, your attention drifts and you go into autopilot mode. If you're making mistakes 30% of the time or more, you'll probably get frustrated and want to stop trying.

Eighty-five percent accuracy seems to be the sweet spot at which we're agitated enough to pay close attention, and rewarded enough that it feels worth the effort. Is it comfortable to be making that many mistakes? Not for most of us. We want our work life—heck, we want life in general—to be one in which we're making as few mistakes as possible.

But discomfort and uncertainty spur learning. Neuroscientists at Yale who work with monkeys found that their brains learned very little when there was certainty.[5] The learning centers of the monkeys' brains, particularly their prefrontal cortexes, basically turned off once the monkeys could predict what was coming next. As soon as the researchers threw in some uncertainty, however, and the monkeys started making mistakes 15–20% of the time, those learning centers lit up and their prefrontal cortexes began working hard to predict what would happen under what circumstances.

So the next time you make a mistake, take a deep breath. You might be out of your comfort zone, but you're in your learning zone.

Neuroscience of Mistakes and Mindset

Neuroscientists are finding that if you want to make fewer mistakes, one of the most important things you can do is to adopt a growth mindset about the skill or ability that's giving you so much trouble. If you have a growth mindset, your brain reacts differently after you trip up.

Some of the most interesting brain research on mistakes has been done with EEGs, which we first mentioned in chapter 1. An EEG records the brain's spontaneous electrical activity and creates a spiky graph that indicates whether there is a lot of brain activity (big spikes) or very little brain activity (small spikes). You don't feel anything during an EEG—the electrodes are recording an electrical signal, not sending one. Like every technology, EEGs have their pros and cons. EEGs are not very good at measuring precisely *where* the brain is active because the electrodes are basically sitting on your scalp and your skull interferes with figuring out exactly where the electrical activity is coming from. But an EEG is incredibly precise at measuring *when* the spikes in activity occur and can pinpoint within milliseconds when something happened in the brain.

Neuroscientists have used EEGs to discover what happens immediately after someone makes a mistake. Participants put on the electrode cap and then the researcher asks them to solve various problems on a computer. The experiments are designed so adults perform well most of the time but make occasional mistakes. (No one wants to make mistakes every time—that makes them just want to give up.) When a participant makes a mistake, the computer records exactly when it was made and what kind of electrical activity the brain performed at that moment. The graph that's created is called an ERP (or event-related potential) and is a very sensitive recording of when and how strongly the brain reacted to the mistake.

The ERP data indicates that mindset matters, though not at first. The first electrical spike the instant after you make a mistake looks the same for all of us, regardless of our mindset.[6] It's what neuroscientists affectionately call the "Oh, crap" moment. It's the moment when you realize you've made a mistake, and you feel discomfort. No matter how resilient you are or how growth-oriented your mindset, you and your brain will have a brief cringe moment when you realize you've screwed up.

But what happens after that reflexive cringe? The brain's second electrical spike, the spike following the "Oh, crap" moment, is larger for people with a growth mindset than for people with a fixed mindset.[7] That bigger spike is basically the brain's way of telling itself to "focus up." If you have a growth mindset, the ERP indicates that you ratchet up your attention for additional input. Mistakes basically trigger your brain into taking in more information when you have a growth mindset— what *was* the right answer?—so you can learn and not make that mistake again. And that boost in alertness works. People who have a growth mindset (and who have that second bigger spike) make fewer mistakes later.

Is anything interesting happening for the people with a fixed mindset or are they just momentarily zoning out? ERPs for people with a fixed mindset indicate that they aren't zoning out. Instead, they have an electrical signal that correlates with being concerned about proving their abilities to other people.[8] So individuals with a fixed mindset are, whether they realize it or not, having a continued emotional reaction beyond the initial "Oh, crap" moment. Instead of being triggered into learning mode, they worry that they're not proving how smart they are. And, not surprisingly, because they're focused more on their reputation and less on their learning, they continue to make more mistakes during the experiment than individuals with a growth mindset. So, with a fixed mindset, you feel worse and learn less.

One fascinating thing that brain scientists have discovered is that these changes in attention and emotions are so fast they can't be consciously driven. For individuals with a growth mindset, they're not consciously thinking *Oops, I should pay more attention now because I just made a mistake* and then reorienting their attention. Reorienting actually happens relatively instantaneously and automatically, outside of consciousness. These changes in electrical activity occur in the first 200 milliseconds after you make a mistake (which is only ⅕ of a second after you realize you clicked on the wrong Zoom button), and depending on your mindset, your brain is either paying more attention so that you can learn, or it's launching itself into regret.

It might be frustrating to hear the response is this immediate and outside your conscious control. But there is a silver lining: Once you've done the work to adopt a growth mindset, you don't have to work hard every time you make a mistake. Your brain will snap into learning mode automatically.

It's Not All or Nothing

Before we dive into strategies for adopting a growth mindset, I have a reassurance to offer. Journalists often describe mindsets as though they're binary, suggesting you either have a strict fixed mindset or a strict growth mindset, with nothing in between. That view is outdated and oversimplified. The reality, as Carol Dweck and others have more recently observed, is that mindsets lie along a continuum, with a fixed mindset at one end and a growth mindset at the other.[9] Some people might be deeply invested in a fixed mindset about an ability (e.g., "I'll always be tone-deaf" or "I'll never be leadership material"), and sit at the farthest end of the mindset continuum, but most people lie somewhere in between the two extremes.

I raise the point about mindsets being along a continuum because researchers find that the size of the second ERP spike is positively correlated with a person's growth mindset score.[10] The higher your score, the more growth-oriented your mindset is and the more your brain reacts in a way that amplifies learning. So wherever you are along that continuum, the goal is to move toward having a stronger growth mindset so that your mistakes help you pay more attention.

What Works

If reading this chapter has convinced you that your mindset is more fixed than you'd like it to be, there's hope. It sounds counterintuitive, but fixed mindsets aren't technically fixed. You can always move toward a stronger growth mindset. The most effective strategies for changing mindsets involve three steps: (1) read, (2) identify actions, and (3) write.[11] Let's look at all three.

1. Read About How Your Brain Changes

You've already taken the first step by starting to read this chapter and, more generally, by starting to read this book.

Researchers find that a crucial step to changing your mindset is to learn how the brain can rewire itself throughout adulthood. That might contradict what you once learned. You might have been taught in school that the brain undergoes all its development during critical periods in childhood, but the research supporting that belief was largely based on animal models. More recent studies using neuroimaging with humans reveal that yes, most of your brain development took place in the first twenty years of your life, but adults still have some neuroplasticity, which, as we learned in chapter 6, is the brain's ability to form new connections based on new stimuli and experiences.[12]

When people learn that their brain is malleable, they begin thinking differently about their potential. Reading is a key step in changing your mindset.

So that's the first thing you should do: keep reading this book. You can also do occasional searches online for "how to rewire your brain" or "strategies for neuroplasticity" to stay abreast of best practices and new research. By reading more about neuroplasticity, you'll instill a firm belief that you're not stuck with the brain you were born with. Heck, you're not even stuck with the brain you had when you picked up this book!

One caution, however, when you're doing your research: Avoid searching for "brain training." Apps associated with that phrase are big moneymakers, but they're not big brain-*builders*. There is one exception, however, which I described back in chapter 1, called BrainHQ, that research reveals is worth considering.

Many mindset researchers like to say that your brain is like a muscle. With effort, it can grow stronger and smarter. The problem with that analogy, though, is that with a muscle, you can do the same exercise many times, such as pull-ups, and that repetition alone will make the muscle grow, whereas with your brain, you need to vary the exercises to see growth. New stimuli, new challenges, and new mistakes. That's what makes your brain change.

So the next time you make a mistake, instead of berating yourself or making excuses, take a deep breath and remind yourself that it's an opportunity. As long as you stay engaged and keep trying, mistakes are a chance to learn and improve your brain.

2. Identify Baby Steps and Track Progress

The next thing you can do to change your mindset is to identify actions you can take to enhance your potential. Pick an area of your life in which you want to grow your potential. Then intentionally revise your language around that behavior or skillset. Don't say, "You can't teach an old dog new tricks." Instead, try "It's never too late to learn." You want to start saying these things internally to yourself as well as out loud to others. Personally, I tend to have a fixed mindset when it comes to any sport involving a ball. (I was so terrible at ball sports growing up that my father gave up playing catch with me.) Two things that I've learned to say, both to myself and to others, are, "I want to be in the learning room, not the proving room" and "I'm not good at _____ yet." That little word at the end is a game changer.

Once you've identified an area in which you want to grow your potential and you've started changing your language, think about what other steps you could take to improve your comfort level or skillset. What's one small thing you could do? For instance, you could put an appointment on your calendar to spend 15 minutes once a week working on that skill. Then note your progress. Keep a running tally of when you use the new skills you're learning and how often you're making old mistakes vs. new mistakes. It could be as simple as a little check mark each day under "old mistake" or "new mistake." You'll probably keep making new mistakes because you're trying new things, but you'll discover that you stop making the old ones. Seeing your progress is powerful.

Here's where a habit-tracking app comes in handy. You can pick a skill you want to work on, then record your progress in the app. There are dozens out there from which to choose. Habitica is popular because it's a "gamified" habit tracker, with stylized graphics that will take you back to video games in the late 1990s. The Way of Life app is particularly easy to use and customize.

After writing this chapter, I decided to take up pickleball, which I've been wanting to learn for several years now. It's not a skill I need at work, but it's an accessible way for me to change one of my most fixed mindsets about myself. First, I changed my language, then I tracked in my journal how much time I played each week and how many mistakes I made in an hour. I went from at least 30 mistakes in my first hourlong lesson (I stopped counting at 30) to only about a dozen mistakes my third time on the court! Of course, I'd rather I didn't make any, but I'll embrace the brain change.

3. Write to Convince Others

The clincher for changing an individual's mindset is to have them write a persuasive appeal to someone else. It's called the "saying is believing" exercise.[13] Here's how it works in the lab: People read a scientific article about how a person's brain can change with experience (the first step above). They learn how their abilities are malleable and how they can grow their potential with focused practice. When they're done reading, they write a letter to someone else who might be struggling with the same challenges, explaining what they've learned about the brain and their potential, what the other person should do when they are struggling, and how they shouldn't give up.

Why is this activity effective at helping someone move from a fixed mindset to a more growth-oriented one? When you write such a letter, you're creating cognitive dissonance because you're writing something that contradicts a deeply held belief. You believe an ability, such as your technical competence, is set like concrete, but you just tried to persuade another person that this ability is more like Play-Doh and can be shaped with effort. Given the contradiction in what you wrote and what you think, something needs to give—and often it will be the way you think.

To make this approach work for you, you could write an email to a friend capturing what you've learned, but if you're active on social media, you've got the perfect platform. Over the next few weeks, post a few times to LinkedIn, Facebook, WhatsApp, or whatever social media platform you prefer, each time capturing a lesson you've learned about how the brain is malleable and how your thinking about your potential is changing. You could even ask your friends to weigh in: In what areas of their life have they gone from embarrassingly bad to pleasantly middle-of-the-pack? Or even impressively skilled?

You could just tell a friend what you've learned (this technique, after all, is called "saying is believing"), but researchers find that writing it out is a bit more effective at changing a mindset than talking it out. (It should probably be called the "writing is believing" exercise, but that's much less catchy.)

Unlike some other practices in this book, you probably won't see immediate changes in your mindset after a single sitting. If you've been saying for years that you're a klutz at technology, that public speaking "isn't your thing," or that you're terrible at names, that fixed mindset is going to take some time to undo. Be patient. At first the effort will be conscious and you'll have to catch yourself when you fall into old patterns of thinking. You'll have to amend your beliefs, saying "No, with effort and extra attention, I will improve." You'll get there.

But Wait, There's More

This chapter is unusual in that it takes one core principle— develop a growth mindset—and breaks it into three strategies for making fewer mistakes. Most chapters have at least two core principles, because, as I noted in the Introduction, I like to give you options. And personally, I love flexibility and choice. But developing a growth mindset is so powerful and affects so many areas of your life, I'm willing to bet everything on this one core approach.

Admittedly, you're probably motivated to change your mindset so that you'll make fewer mistakes, but changing your mindset can change your *self* in a number of other surprising ways. Researchers find that adults with a growth mindset enjoy a peace of mind at work that their colleagues with fixed mindsets don't.[14] Chances are you'll feel less defensive when you receive feedback from your supervisor. (Wouldn't it be nice to feel less threatened in those one-on-ones?) Research also suggests that you'll feel more creative with your work and more willing to try new ideas that you once thought were too risky, in part because you realize that it's okay to take risks and make mistakes now and then.

In what may be the most unexpected twist, with a growth mindset you'll also be more likely to confront either prejudiced comments you hear or biased behaviors you see at work.[15] Because you increasingly believe people can grow and change and you realize your colleagues aren't necessarily set in their ways, instead of being a silent bystander who just grits their teeth when you see someone behaving badly, you'll be more willing to find ways to speak up.

So find an area of your work in which you have a relatively fixed mindset and try to grow it. You'll benefit in many ways and your team will too.

TRY THIS
Your "Make Fewer Mistakes" Toolkit

The essential strategy here is to move from having a relatively fixed "I'm just not good at this" mindset about something you don't do well, to having a growth mindset and believing "I'm not good at this *yet*." There are three key steps you can take.

▶ **Keep reading about how the brain changes with experience.** In addition to reading this book, find articles online using search terms such as "how to rewire your brain" and "strategies for neuroplasticity."

▶ **Identify baby steps.** Think of phrases you'll say to yourself and others when you make a mistake, such as "I'm not quite good at _____ yet" and track your number of mistakes so that you can see improvement over time, however gradual.

▶ **Persuade someone else.** Write an email to a friend or write a post on social media for anyone who might be struggling with the same challenges and explain your own struggles and improvement, how they shouldn't give up, and what you've discovered about how the brain changes by learning from your mistakes.

BE BETTER TO OTHERS

CHAPTER 8

Relate More

You start off the day in a good mood. But once you look at your meeting schedule, your whole body slumps. You think to yourself, "Seriously? It's going to be one person after another whining about their problems." You resign yourself to putting on your understanding face and getting through it, but you know these back-to-back meetings are going to take a toll.

We've all been there, but what if, instead, you looked at your calendar and thought, "Interesting—I've got an opportunity to take care of so many people. What an incredible day this is going to be!"

The paragraph above might seem like foolishly optimistic thinking (or perhaps it sounds so saccharine that you threw up a little in your mouth just reading it). I hear you. Before I started researching this book, I would have thought such an attitude was either unrealistic or required some level of Dalai Lama–like empathy I could never muster.

I'm here to tell you that it is realistic and it doesn't require the empathy of a global spiritual leader. But empathy is a good place to start. Empathy has become such an HR buzzword that companies are spending millions of pounds training leaders how to show more of it. Surprisingly, neuroscientists have some answers that many well-intentioned trainers miss. In this chapter, I'll review cutting-edge research findings about the neuroscience of empathy, how empathy can be helpful, and how to increase it, but we'll also explore why you might be better off focusing on something *other* than empathy.

Empathy Comes in Multiple Flavors

You might be wondering what, exactly, empathy is. You're not alone. As of 2016, psychologists had 43 (!) different definitions for empathy, and more have been added since then.[1] Confusion abounds.

What most experts agree upon today is that there are at least two kinds of empathy, and both are about understanding another person's emotions. One kind is emotional or affective empathy, which is when you feel someone else's joy or pain and share their emotional experience. The other kind is cognitive empathy, commonly referred to as perspective taking, which is when you perceive that someone else is feeling joy or pain and you're able to infer how that might affect them.[2]

To clarify the difference between these two kinds of empathy, imagine that a colleague didn't get a promotion she expected. If you're experiencing emotional empathy, you feel her sense of shock, distrust, and intense disappointment, maybe rising anger, as she starts telling you about it. You might feel ill as she describes it or cry along with her. Or, if you're feeling cognitive empathy, you perceive that she's in pain and deeply disappointed but don't feel any of those emotions yourself. You might be thinking about what those emotions mean for her, and perhaps predict that she won't be able to focus this afternoon. Or you might run through both kinds of empathy, first feeling along with her, then moving into thinking and predicting mode. Put simply, emotional empathy is feeling, cognitive empathy is thinking.

People vary widely in terms of which kind of empathy they experience and with whom they experience it. You might have more emotional empathy for your close friends and family members but have more cognitive empathy for your colleagues. That might mean you can see how someone feels at work and quickly move into problem-solving mode, but because you're not feeling their pain right there with them, they might be bothered that you're not more upset on their behalf. Alternatively, you could be someone who has a high degree of emotional empathy for most people, even for those with whom you are not close. Personally, that's my challenge and I have, for better or worse, been known to start crying a second or two before the person across from me starts crying, even if we hardly know each other. Or you could face the challenge of not experiencing either kind of empathy readily, which means you have to work harder to decipher someone else's emotional cues and how those emotions might be affecting them. You notice your manager's brow is furrowed, but is that concentration, confusion, or anger? And does that mean you should explain further or shut up?

There is no wrong or right way to be, but you'll face different challenges depending on how much emotional and cognitive empathy you experience.

Show Me You Care and I'll Stay

It's obvious why you'd want to have empathy for your children or your spouse, but it might be less obvious why it's important at work. Empathy at work matters because few of us leave our home lives completely at home. Consider your personal life for a moment and how it affects your workday. If you're up all weekend with a crying infant, you'll feel it Monday. If your mom has texted you three times today about your dad's worsening dementia, you'll likely find it hard to focus. Or maybe you're having your own health problems, so your motivation is zapped.

If your colleagues or your managers empathize with you as you go through these trials and tribulations, it helps. A lot. Researchers find that when employees see their leaders demonstrating concern and understanding for the tough circumstances they're facing outside of work, those employees dig deep and work harder.[3] Employees at high-empathy companies innovate more and feel more engaged, despite whatever stresses are looming at home, *and* they feel more loyal, saying they're less inclined to leave. As a result, some business analysts have gone so far as to say "empathy is the most important leadership skill" in workplaces today.[4]

Most of us assume that if someone cares and feels more, they'll do more to help. We assume an empathetic manager will be more willing to listen to what's troubling us than an unempathetic manager and will be more understanding if we want to submit our portion of a project Sunday night instead of Friday morning. We assume empathy is key to being supported at work. (We'll come back to this assumption in a moment.)

What Works

1. Find Your Motive

There is good news and bad news when it comes to experiencing higher levels of empathy, though. Let's start with the good. If you've received the feedback that you need to "care more" or "take other people's perspectives" and you're feeling pressure to have more empathy, there is hope. Most people can increase their emotional and cognitive empathy with some effort, so chances are you're not stuck with your current empathy levels. (I say "most people" because certain conditions severely disrupt empathy. Narcissism, autism, and bipolar disorder are some of the conditions that lead to empathy deficits, much to the dismay of everyone around them.)

Assuming you are like most people, how do you increase your empathy levels? Motivation is key. Researchers have found that if you're highly motivated to empathize, you can.[5] When you're motivated, you'll find yourself orienting your attention to different cues by paying more attention to someone's emotional nail-biting, hunched shoulders, or wiggling foot, and voilà, you're right there with them. In fact, researchers find that when most people aren't feeling empathy, it's not because they're *unable* to feel empathy, it's typically because they're not sufficiently motivated to empathize (or more accurately, they're feeling conflicting motives).[6]

You could tap into several possible motives to ramp up your empathy.[7] One might be a desire to connect and belong. Or you could be driven to empathize with someone who is struggling because you want to feel closer to them or because you want to be part of a team that cares. You might also be driven to empathize because you see it as a personally desirable quality and it makes you feel good about yourself. If you decided that "being kinder and more caring" is one of your priorities for the year, you'll be galvanized to empathize more. Lastly, you'll be more motivated to empathize with people who are like you. If you focus on your similarities to someone else (and less on your differences), you'll find it easier to understand or even feel their feelings.

If one of these motives resonates with you, take a few minutes to think about it before going into a meeting in which you want to empathize. Is your goal to connect and belong, to become a better person, or to see what you have in common? All are worthwhile motives that will help you lean in to empathy.

What's tricky here is that on any given day, you might be highly demotivated to empathize. And that's the bad news— motives work in both directions, dialing empathy up or down. Let's go back to that moment when you're looking at your schedule and realizing you have several meetings with people who are struggling. If you're thinking, *How am I going to get through this?* you'll be motivated *not* to empathize—you'll be focused on conserving enough energy to survive the onslaught. Psychologists call it "anticipated exhaustion" and in day-to-day life, it's one of the main drivers not to empathize.[8] You know you're going to be mentally and emotionally taxed if you care too deeply or do a lot of perspective taking, so conscious or not, you brace yourself and don't empathize. It's a self-preservation tactic, and a savvy one at that.

DOES AGE MATTER?

If you're in your sixties, you may find yourself thinking, more and more often, "That behavior seems crazy to me." Maybe it's all the piercings of the twenty-two-year-old who serves you coffee, or maybe it's the yard full of old appliances that you see when you drive through your neighborhood. You once thought of yourself as accepting people from all walks of life, but now you find yourself increasingly puzzled (and sometimes annoyed) by what feels like other people's bizarre choices.

Aging will do that to you. Cognitive empathy, the ability to take someone else's perspective, typically changes over the course of our lives. It increases from our teenage years until our late thirties or early forties, peaks, and then begins to drop.[9] That can make it harder than ever to understand why a colleague who is twenty years younger does what they do, but it's not just the age difference that poses a challenge. You might be equally baffled by a friend's ridiculous romantic choices—they're moving in with someone after dating for only three months?—even though you and your friend are the same age. Perspective-taking probably isn't as easy as it once was.

Two things can help. Education is one. The more educated a person is, the easier it is for them to practice perspective-taking as they age.[10] But since you might not be interested in going back to college at age sixty, thankfully you have another option: work on your motivations. Older adults who are highly motivated to have cognitive empathy are able to do perspective-taking as skillfully as their much younger peers.[11]

So if you find yourself thinking, "I really need to have more empathy," figure out which of those motives described earlier resonates most with you. Or create your motive. Maybe you want your legacy to be that you cared about other people. Maybe you want the pleasure of having more friends.

How about emotional empathy? Research reveals that emotional empathy remains steady or, in many cases, increases with age.[12] Your ability to feel what other people are feeling is just as high, maybe higher, than it ever was, especially when you've just interacted with a person or watched them go through something difficult.

If you're struggling to tap into your empathy at work, invite the person whose actions baffle you out for coffee. It might be hard to understand their reasoning based on what they did in a team meeting, but if you get to know them better, you can tap into your emotional empathy. You could try saying, "I'm curious about your perspective on the X project. I'm hoping to learn from you. Would you mind sharing your thoughts?"

Your Pain Has a Volume Knob

Motives are tricky. If you feel strong emotional empathy, you could be demotivated by the biggest deterrent of all: pain and distress. The neuroscience is clear on this point. Someone else's pain and distress can quickly become your pain and distress. To understand how that happens, we need to take a quick look at the anterior insula, a brain region crucial for empathy.

I find the anterior insula fascinating. You actually have two, one on each side of your brain, an inch or two in from your ear. Insula means "island" in Latin, and the anterior insula is a small oval island that has many connections to pain receptors across your body. The anterior insula plays a key role in perceiving pain and evaluating what hurts most. Let's say you're on your morning run and you fall, scraping yourself up like a seven-year-old. It's your anterior insula that's taking in all those pain signals and telling you, "Your hand! Look at your bloody hand!" Your left knee hurts too and may even be bleeding more, but you'll check that later because your anterior insula is prioritizing your hand.

What's fascinating about the anterior insula is that it determines how much something hurts.[13] I always thought that if there's a pain signal coming from a nerve, you'll feel pain, and the intensity of that pain will be determined objectively by how strongly that nerve is firing. Perhaps, like me, you can picture a diagram you saw in Introduction to Biology in which a hand reaches out to touch a hot flame, a pain signal travels to the spinal cord, a new signal travels to the muscles in the arm, and the person reflexively withdraws their hand. That's part of the story, but far from the whole.

Let's imagine that you've been asked to extend your hand and hold it above a candle flame. If you've been told, "This is an entirely safe distance to hold your hand from a flame for several seconds and others have been able to do this just fine," your anterior insula will stay quiet as your hand hovers above the flame. You'll still withdraw your hand once it hurts, but because your anterior insula isn't agitated, you won't feel nearly as much pain. You might even say, "That wasn't so bad after all."

But if you've been told, "Holding your hand this close to a flame can be painful and others have found this quite uncomfortable," your anterior insula lights up before you've even started to move. It's sending a threat message. As a result, even though the nerves in your palm receive the exact same pain signal from the heat of the flame, the anterior insula has amplified the signal, and you'll feel much more pain. In a sense, the anterior insula sets the volume knob on the pain you feel, determining whether that sensation is a mild 3 out of 10 on the pain scale or a searing 7.

The Pain Is Real (Even If It's Someone Else's)

That might seem like a bit of a digression, but it's important because the anterior insula doesn't just turn up the volume when something painful happens to you. It can also turn on the pain signal when something painful happens to *other* people, especially if it's someone you're motivated to care about. Neuro-imaging studies reveal that the anterior insula becomes highly active when you watch a friend or a loved one in pain.[14] (I'm not a parent, but I've heard many mothers and fathers say how deeply it hurts them when their child is in pain.) Even though the pain receptors in your body aren't sending any signals, your anterior insula becomes active and you hurt too. What seems like "your" pain center is, in actuality, "our" pain center. Empathy joins us together, and shared pain can become real pain.

As a result, empathy can hurt. For example, in the case of your colleague who just learned that they didn't get the promotion they'd hoped for, you're not just imagining pain, your brain is "feeling" the pain in much the same way it would feel if *you* didn't get that promotion. Psychologists call it "empathic distress."

It sounds like a lovely and generous quirk of human physiology, but empathic distress can actually be a real problem. Instead of empathy leading you to help someone else, empathy can shut you down. If the hurt is too great, you retreat to deal with your own distress and don't have the resources to provide support to others. Research bears this out. Earlier, I said the assumption was that empathy would lead an individual to help others, and sometimes it does, but empathy, especially the emotional kind, can also lead individuals to retreat because now they need to deal with their own emotionally upsetting experience.[15]

You might be thinking, well, let's avoid emotional empathy and focus on increasing cognitive empathy at work instead. And you'd be right—cognitive empathy doesn't activate the anterior insula, so it doesn't trigger vicarious discomfort in the same way that emotional empathy does.

The problem is that cognitive empathy is very mentally taxing. It takes a lot of mental work to take another person's perspective, especially if you're doing it with one person after another, and research shows most people don't want to work that hard.[16] You might do highly effective perspective-taking with two or three people, thinking through all the nuanced ways they are experiencing their situation, but then you may feel so tired that you lose your motivation and glaze over for the rest of your meetings.

If you only have to interact with two or three people in a given day at work, cognitive empathy and perspective-taking can be your go-to tool for taking care of others. Or, if you find that you're not all that taxed by empathizing, either emotionally or mentally, then use the motives listed above to dial up your empathy when you need it. But if you're like many of us and have too many people tugging at you or are quickly exhausted when you empathize with even a few people, you need another solution.

Empathy vs. Compassion

Thankfully, there is one. Say hello to compassion. You might use the words "empathy" and "compassion" interchangeably, but psychologists make a distinction (and, as you'll see, your brain does too).

Empathy, as we've seen, is the ability either to think through someone else's emotions or to vicariously feel those emotions. For this reason, it is often described as "feeling *with* you."

Compassion, in contrast, is feeling *for* you. It has been defined as a "feeling of concern accompanied by a motivation to help."[17] Whereas empathy is feeling or understanding that someone is sad, compassion is feeling a need to *act* because you see someone is sad.

Empathy is akin to walking in someone else's shoes, such that you're seeing or feeling the world as they do, whereas compassion is akin to observing that someone has crummy shoes and offering to call a cobbler. The two concepts are related, so the distinction might seem fuzzy. In both cases you're attending to someone else's feelings, but with compassion, you're keeping a little more emotional distance, plus you're more focused on what you can do to help.

An Energizing Spa for Your Brain

Your first reaction might be that this is silly semantics, but your brain reacts to empathy training very differently from compassion training. In one clever study, a team of European researchers first gave individuals a classic form of empathy training, specifically, increasing participants' emotional empathy for a stranger who was struggling.[18] When the scientists scanned the brains of these individuals trained in empathy, they found there was increased activity in the anterior insula. In essence, their brains were saying "this is important and it hurts." They'd become more sensitive to another person's struggles, true, but in a way that was hard on them. Compared to individuals who hadn't been trained to feel empathy, those with training felt stronger negative emotions when they saw someone in distress, whether they saw someone in a little distress or a lot of distress.

Then the researchers gave these same individuals compassion training (we'll describe what compassion training looks like below). And compassion training made a world of difference. First, compassion training eliminated empathic distress and set participants back to an emotional baseline. Now when they saw someone struggling, they were still bothered, but not with the same intensity that they'd had after the empathy training. With compassion training, individuals seemed to be better able to regulate their emotions and soothe the difficult bits. Second, the people with compassion training actually felt an uptick in positive emotions when they saw someone struggling. You read that right. This wasn't schadenfreude—a pleasure from someone else's misfortune—but rather a sense of warmth and caring toward the person who was struggling. And they didn't know the people who were struggling—they were just watching a stranger in a video.

Lastly, the compassion training was like an energizing spa day for their brains. Whereas empathy training agitated the anterior insula, the compassion training quieted it down and activated the ventral striatum instead. We first encountered the ventral striatum in chapter 3. As we've seen, the ventral striatum is associated with feeling rewarded and motivated. The ventral striatum often becomes activated when you're figuring out what you should do and convincing yourself it's worth doing. Part of the ventral striatum also tends to be highly activated when we're doing something to help others, especially when we're doing something to help people we don't know well.[19]

It's interesting that this brain area was activated during compassion training, not empathy training. Whereas empathy training increased an observer's distress and triggered their brains to go into "this really hurts" mode, compassion training reversed that problem and instead made individuals feel a sense of warmth and kindness for this stranger who was hurting, *and* it created a strong motivation to help the stranger.

And help they do. Whereas empathy training sometimes leads individuals to help and sometimes leads them to retreat, compassion training more consistently leads people to help.[20] It hurts less and you feel more motivated, an energizing combination. If you're hoping to help someone who's struggling while preserving your own positive mental outlook, compassion training is for you.

2. Try Compassion Training and a Different Kind of Meditation

That brings us to our second strategy for relating more to others: Try compassion training. Compassion training typically involves learning how to do something called loving-kindness meditation, which is different from the mindfulness meditation we learned about in chapter 6. (If you thought all meditations were equal, get ready to be enlightened.) Whereas mindfulness meditation involves focusing and refocusing one's attention on the present moment, loving-kindness meditation involves creating a positive feeling that you extend to others.

What does this kind of loving-kindness meditation entail? Although every compassion training or loving-kindness meditation is unique, they generally share a few key components.

- You are guided to visualize a time when you experienced your own suffering.

- You are encouraged to relate to that experience with feelings of warmth and self-care, saying to yourself things such as "May I be safe" or "May I be happy."

- You are then asked to sequentially visualize extending that feeling of caring first to someone you're close to, then to someone with whom you have conflict, then to a neutral person you don't know well (perhaps a coffee shop barista), then to complete strangers and humanity in general, repeating the phrases "May you be safe" and "May you be happy."

These steps typically constitute a single meditation exercise, lasting 10–30 minutes. You practice this meditation several times, learning new skills and gaining new insights each time. A training might be an intensive full day, or it might be 2 hours at a time over several weeks. Expect to invest at least 8 hours to learn and practice the fundamentals of loving-kindness mediation.

You might be thinking that this is a lot more involved than other practices in this book. Is it as easy as taking a supplement every morning or putting on headphones? No. But are there bigger benefits? Yes. Research on self-compassion and loving-kindness meditation reveals that if you practice them regularly, you'll flourish on multiple fronts, from feeling more happy and hopeful to improving something called cardiac vagal tone (which strengthens your gut and your heart). You might even lengthen your telomeres, a change associated with a longer lifespan.[21] Your colleagues at work will benefit immediately, and you'll benefit immediately *and* long after you leave that job.

Look for compassion or self-compassion training courses or loving-kindness meditations online. Either can offer what you're looking for. Despite the different names, both compassion and self-compassion training typically teach you first how to offer compassion toward yourself and then how to extend those feelings toward others.

For the Skeptics

What about all those findings that employees want leaders with high empathy? Employees would probably appreciate high-compassion leaders even more than high-empathy leaders, since compassion spurs more action and helping behaviors. But compassion training isn't common in many workplaces. Can you imagine seeing "Loving-Kindness Meditations for Managers" flyers in your cafeteria? They would probably be quickly covered with snarky comments. Because compassion isn't a buzzword, employees don't know that's what they really want from their managers.

After researching this chapter, I decided I needed to give compassion training a try. I completed an online self-compassion training course with Kristin Neff and Chris Germer, cofounders of the Center for Mindful Self-Compassion and leading researchers in this domain. It was 12 hours, much of it during the workday, which was hard to schedule. But the results? It was nothing short of transformative. Rarely have I participated in a workshop and found myself turning back to what I learned six months later. The primary benefit that I've seen is that it's helped me be gentler with myself in high-stress situations, treating myself as I would a friend. But it's also helped me relate more to colleagues and family members when they're struggling or suffering. As someone who is high in emotional empathy, I can easily become distressed when someone else is upset, which disables me and doesn't empower them. But now that I know some compassion tools, when someone I care about is upset, I find it easier to stay clearheaded and genuinely helpful, rather than being pulled into their pain.

If this still sounds too woo-woo for you, perhaps it helps to know that certain professions have embraced compassion training with great success. Healthcare professionals, for example, have been touting the benefits of compassion training for doctors and nurses because it has benefits both for patients and caregivers. Healthcare workers feel more compassion toward their patients and themselves, less conflict with their colleagues, and less burnout.[22] One research team called it "self-care for the caregivers."[23]

TRY THIS
Your "Relate More" Toolkit

▶ **Find your motive.** If you want to increase your empathy, you need to find a motive that truly energizes you. You could be energized by the desire to connect and belong, by the desire to become a better human being, or by the desire to see what you have in common with someone else. Tap into one of these motives and it should be easier to relate.

▶ **Try compassion training and a different kind of meditation.** Compassion has benefits over empathy, but it's harder to learn and takes more time. Look for either "compassion training" or "self-compassion training" online and learn how to do loving-kindness meditations.

▶ **Does age matter?** Cognitive empathy tends to drop after our late thirties/early forties, but finding your motivation can restore your ability to do perspective-taking. Emotional empathy, on the other hand, tends to increase as we age, so you could try to relate to someone's feelings, perhaps by asking, "How are you feeling about X?"

CHAPTER 9

Be More Fair and Less Biased

Your team is working hard to meet a fast-approaching deadline, an important one you absolutely *must* meet. You get an email from HR about a mandatory two-hour workshop on unconscious bias that you're expected to attend, reminding you that it's this week. *This week?* You're tempted to beg for an exception. You agree the training is important, but couldn't you attend next month, when the pressure is off? You're convinced you'll get more out of the workshop later, when you can focus.

We've all had workshops that we've agreed to months in advance, then later wished we could skip. But don't postpone the training. It feels like the timing couldn't be worse, but as you'll see in this chapter, in many ways, the timing couldn't be better, especially if you're a manager and employees report to you. Right now, when that deadline is breathing down your neck and you're feeling stressed, is when you're most likely to be biased and unfair.

Now who's being unfair?

Perhaps I am. I don't know you, of course, and I don't know how you react under pressure. Maybe you're not feeling anxious or stressed at all. But most of us would be. Before we talk about the ways to combat bias, let's take a quick look at the science of bias to understand when it typically rears its ugly head and how it changes for people under pressure.

People Who Are Just Like You

Definitions first. Simply put, a bias is a preference toward or against something in your environment. A bias can be harmless, like a bias for Coke over Pepsi, or a bias can be harmful, like racial bias that leads to unfair treatment. Think of bias as leaning. When you're biased, you're leaning—either toward or away from—something or someone.

Before we go any further, let's clear up a common misunderstanding. A popular assumption is that bias only occurs when people believe something negative about a group. This isn't true. Let's say a hiring committee that's 75% white is leaning toward hiring a white candidate over a Latino candidate, even though both are qualified for the job. "It's not that we don't like Carlo," they say, "but Kent just feels like a better fit." It could be that what's affecting the hiring committee are some unconscious negative beliefs about Latino individuals, but research reveals that what's probably swaying them even more is their unconscious preference for white individuals.[1] We tend to like people who are like us. It's bias, and psychologists call it "ingroup bias" or "ingroup favoritism." You find yourself perking up and suddenly liking someone more when you learn they attended your college, root for your sports team, or come from your hometown.

Perhaps you've never thought about your ingroup, but it's worth taking a few minutes to identify who that might be. For starters, our ingroup often reflects the basic demographics we'd check on a survey. What's your race or ethnicity, gender, age, and sexual orientation? If you're white, your ingroup would be white. If you're fifty-four, your ingroup might be other people over fifty. (Or if you're fifty-four and most of your company is people under thirty, your ingroup might be anyone over forty because they're the only people who genuinely laugh at your jokes.) Many other factors might be central to your identity, such as your job title or role, your nationality, your political affiliation—basically any area in which you find yourself thinking (though you'd probably never say it) in terms of "us and them." As you can see, you actually have several ingroups or groups in which you feel like you belong. For me, my ingroups include individuals who are female, frugal, between the ages of fifty and sixty, and live in the Pacific Northwest. I tend to feel a kindred spirit with members of those groups, and the more of these boxes you tick, the more my eyes are likely to light up when I meet you. You can make a similar list for yourself.

Now that you know your ingroups, it's easier to identify your outgroups. That isn't a criticism of those groups, just that they are groups to which you don't belong. It doesn't mean you don't like them or respect them or believe they're equals in every way. But your brain automatically categorizes people as "like me" or "different from me." If you're going to be on the lookout for bias creeping into your actions and reactions, you need to be aware of whom you're categorizing as "different." Race and gender are processed in milliseconds, so people of a different race or gender will be in your outgroup.[2] In many professions, your role, training, and title strongly define ingroup and outgroup. I'm an academic, and in universities there are strong distinctions between faculty and staff. In healthcare, lines are drawn between doctors, nurses, and administrators. It may not be pretty (and please don't post it on social media), but take a minute to identify your outgroups.

Fighting Ingroup Favoritism

Ingroup favoritism shows up in many ways, but one concerning pattern is that you'll have more empathy for members of your ingroups and less for those in your outgroups. As you'll recall from chapter 8, empathy is understanding the feelings and mental states of someone else, and in some cases, even sharing those feelings. Researchers find we have greater empathy for people of the same race and, as a result, if we see them as having potential, we want to do more to help them.[3] This is a huge problem in hospitals, where studies reveal that white medical staff prescribe more pain relievers for white patients than for Black patients. But having greater empathy for your ingroup can also lead to major inequities within teams at work.[4] If Jennie, who happens to be the same race as you, asks you to reduce her workload, you're more likely to accomodate her than Juan, who has a different skin color.

Why do we have more empathy for members of our ingroup? Pain signals, believe it or not, are one reason. Neuroscientists have found that our brains are more sensitive to the subtle pain signals on someone else's face when we're looking at someone of our own race, but when we're looking at someone of a different race, we require more obvious signals.[5] When we view someone of the same race, our brain can register *She's in pain* with just the tiniest change in her forehead, eyes, and mouth. But when we view someone of another race, our brain requires much more obvious signs of pain and distress to register *He's in pain, too.*

Let's consider what this means for a team that's putting in long hours to meet a deadline. I happen to be white. If my manager is also white, I only have to be a little visibly distressed for his brain to register that something's wrong, whereas one of my Asian or Middle Eastern colleagues would have to be much more visibly distressed for our manager's brain to notice. He may think, *Therese seems unusually stressed* before I've even said anything, and he may cut me some slack but not extend that courtesy to my colleagues because as far as he can tell, they seem to be doing fine. They have to ask for support; I don't.

To make matters worse, research suggests that you probably feel immediately rewarded when you favor someone in your ingroup. Neuroscientists have used imaging techniques to reveal that the ventral striatum, which is activated when you're feeling rewarded, jumps for joy when you do something that's costly for you but makes life easier for someone in your ingroup.[6] It doesn't literally jump for joy, of course. More accurately, when you show ingroup favoritism, the ventral striatum suddenly becomes activated, which, as we learned in chapters 3 and 8, means you feel rewarded. Help someone in your ingroup and it's like brownies for your brain.

You can see how tempting that might be, especially during a hard week. You give someone who looks like you an extension, and that act of generosity might be costly and create extra work for you, but it also makes you feel like a good human being, a fair and reasonable manager. You'll have a gut feeling that it's right to be generous to a member of your ingroup. And most of us welcome warm, rewarding feelings during a week that's anything but. The problem, though, is that you won't get that same gut feeling or sense of reward if you favor a member of your outgroup. The ventral striatum doesn't automatically do a happy dance when you help a member of your outgroup, so you'll have to do the extra conscious work to reassure yourself that yes, giving that person an extension is the right thing to do. But it won't feel as immediately or as intrinsically reward-ing, which means you'll probably focus more on reasons not to help them.

Stress Makes You Like Teflon to Outsiders' Needs

So far, the news on bias hasn't been good, I know, but it's about to get worse. Fear and anxiety amplify bias. Have you ever noticed that you feel less concerned about other people's needs when you're stressed? A team of psychologists have found that fear and anxiety don't change the empathy you feel toward members of your ingroup, but these unpleasant emotions dramatically reduce the empathy you feel toward members of your outgroup, especially for members of a different race.[7] We become like Teflon; if an outgroup member comes to us with a need while we're feeling anxious or worried, that person's needs slide right off.

This is why bias is more likely when you're stressed. You could be anxious that your team won't meet this looming deadline or that your manager will be frustrated if the work isn't polished. That anxiety will distort your sense of who you want to help. You feel fine helping Jennie because empathy for your ingroup hasn't diminished, but you're less likely than ever to help Juan. The stress has caused your empathy for him to plummet.

What's fascinating is that researchers find that if they block the brain's stress and anxiety signals, then empathy for the outgroup is quickly restored. The science on this is still relatively new, but one team of neuroscientists found that if they used drugs to block feelings of anxiety, then empathy remained high for one's outgroup.[8] You don't look like me, but hey, I'm relaxed so I still want to help you.

What's the takeaway? Fear and anxiety make you more biased. So if you're terribly worried that your team won't meet this deadline, you'll behave in biased ways that aren't like you.

Not Your Destiny

Some people may read this chapter and think, "If my brain is wired for bias, there's nothing I can do." They'll simply shrug and go on operating as they have, believing that bias is an inevitable part of being human.

I hope you're not one of them. If you've learned anything from this book, it should be that you *can* change your brain. As we've seen in nearly every chapter, your brain isn't like those parts of your body that are irrevocably determined by your genes. If you're born with brown eyes, you'll have brown eyes for life (blue contact lenses notwithstanding). But if you're born inclined toward bias—and we all are—you can overcome that automatic reaction and instead respond with empathy toward members of your outgroups. Engage in the right activities and you can reduce the knee-jerk tendency to favor your ingroup. Let's look at which strategies are most effective at overcoming bias.

WHAT DOESN'T WORK (AT LEAST NOT WELL)

Psychologists have tried many, many strategies for reducing bias and prejudice, and as a result, we know quite a bit about which strategies are more effective. There are two approaches that sound promising and might even feel good while you're doing them, but according to the research are less worthwhile than other approaches described in this chapter.[9] (I'm not saying these approaches never work. But if you're pressed for time, you'd be wise to focus your precious energy on other, more reliable strategies.)

The first strategy that doesn't work as well as one might hope is mere face-to-face contact or exposure to members of your outgroups. Simply working with someone of a different race or ethnicity, for example, doesn't change your thinking about the larger group to which that person belongs. If you like that person, you might think of them as an exception to the rule, and if you don't like that person, you might (consciously or unconsciously) use it as evidence to reinforce a stereotype. I've seen the latter happen in tech when a male manager has his first female software engineer on his team. If she struggles in her first few months, he uses it to reinforce his stereotype that men make superior programmers. He invests less energy in her, and she never gets a chance to impress him.

The other strategy that seems to be less effective at reducing bias is having conversations about bias with other members of your ingroup. Perhaps you're white and you talk with your other white friends about racism. Those dialogues may be validating and insightful, perhaps even guilt-relieving (and that can be valuable in its own right), but unless the conversation includes some of the tactics discussed elsewhere in this chapter, the research suggests it probably doesn't curb bias as much as you'd like. It's less clear why this strategy falls short, but it may be that you spend more time reassuring each other and less time challenging existing beliefs and behaviors.

What Works

Thankfully, there are effective strategies for curbing ingroup bias.

1. Learn When You're Likely to Be Biased— and Create an Action Plan

Neuroscientists have discovered there are at least two key brain areas involved in reducing bias and prejudice, one brain region for detecting that you're about to react in a biased way, and

another brain region for saying, "Nope, let's do this instead."
Your goal is to train both brain areas.

The first brain region, the one that detects when you're about
to be biased, is the dorsal anterior cingulate cortex (dACC). The
dACC lies deep inside your brain, toward the front. (For those
who remember the corpus callosum from their Introductory Psychology class—the thick bundle of fibers that connects the two
hemispheres of the brain—the dACC rests on top of it.) Think
of the dACC as your vigilance area, like a security room in your
brain, with a wall full of TV monitors keeping tabs on certain
other brain areas.[10] It's aware of your current urgent goals, such
as "Need to meet this important deadline," as well as your less
urgent, everyday goals, like "Need to treat everyone equitably."
The dACC is busy doing a lot of error monitoring to make
sure your behaviors align with your goals. So, one way you can
reduce bias is to train your dACC to notice those situations
when you're inclined to favor your ingroups.

How can you discover when you're likely to favor one of your
ingroups? One way is to take the Implicit Associations Test, often
called the "IAT." You can find it online by searching for it by
name or by going to implicit.harvard.edu. Developed by a team
of researchers, the IAT is actually a set of several online tests that
help you determine where you have the strongest unconscious
bias. It's an eye-opener. Are you likely to favor thin people over
fat people? Do you have an unconscious bias that men should be
at work while women should be at home? Perhaps you answered
a vigorous "no" to both of those, and that's the conscious part of
your brain talking. The IAT asks, "But what are the unconscious
parts of your brain saying?" You might be surprised. The IAT
will reveal parts of your life where your unconscious brain has a
different view than your conscious brain. Each test takes about

10–15 minutes and you'll receive a score, indicating whether you have more or less bias than others who have taken the test.

It takes courage and curiosity to take the IAT, and if you're like me, you'll retake some tests immediately because you'll be thinking, "That can't be right. I don't think that." But if you're committed to being fair, the IAT will help you identify where your unconscious mind sees people differently from your conscious mind. It will help you discover your blind spots.

Of course, simply knowing your blind spots isn't enough. You also want to ensure that your behavior aligns more with your conscious mind, less with your biased unconscious mind. That brings us to the second brain area.

The second brain area that's involved in reducing bias and prejudice is the dorsolateral prefrontal cortex on the left side. The dorsolateral prefrontal cortex, broadly speaking, is generally your "better self" center. Instead of sinking to a self you might be ashamed of, the dorsolateral prefrontal cortex helps you rise to a self you might be proud of. The right and left sides of the dorsolateral prefrontal cortex do different things. We learned about the right dorsolateral prefrontal cortex in chapter 4 when we learned about how to do something hard now and delay satisfaction until later. Essentially, the right side stops you from taking the easy road.

Your left dorsolateral prefrontal cortex serves a slightly different function—it can prevent you from behaving in biased ways. The left dorsolateral prefrontal cortex becomes active when you're restraining yourself from being sexist or racist and deciding how you'd rather act instead.[11] It helps you find a better alternative. So if Juan asks for an extension and you're about to say no, it's your dACC that chirps, "Wait a second. You said yes when Jennie asked this morning. Are you being fair?" and it's your left dorsolateral prefrontal cortex that makes you feel guilty. Thankfully, the left dorsolateral prefrontal cortex does more than make you feel bad. It also chimes in with, "Let's figure out a fairer or better solution" and starts working to find one.

The challenge here is that your brain is inherently lazy and will default to whatever is easiest. You might cringe momentarily, then justify your ingroup favoritism to yourself, perhaps thinking, *Jennie really seems like she's struggling,* and she probably does seem that way (to you at least), but again, that's your bias noticing her pain signals and not Juan's.

So you want to provide your left dorsolateral prefrontal cortex with a fair action plan that you've already thought through, a plan that doesn't favor your ingroup, so you can grab that plan, rather than the biased action plan that's so readily available. For instance, in the case of giving people extensions, instead of basing it on how stressed someone appears, create a more objective metric. Perhaps anyone who worked overtime on the last deadline gets offered an extension on this one. You'll have to decide what's fair and reasonable, just don't trust your gut when it comes to these kinds of people decisions. Your gut feeling, like it or not, is most likely unconscious bias.

So, to recap: The neuroscience teaches us that there are two valuable ways to reduce bias: Train your dACC to notice when you're inclined to favor your ingroup, and give your left dorsolateral prefrontal cortex equitable action plans that you can fall back on when those situations arise.

2. Attend Unconscious Bias Workshops

You might be thinking that with your big deadline looming, you don't have time to take the IAT or to identify fair, unbiased action plans, at least not right now. Here's where unconscious bias training can assist you. Sure, the training takes time, but it can quickly uncover fair and unbiased actions for you, freeing your mind up to focus on that deadline. Whoever is offering the training, be it HR or the DEI team, they've hopefully identified specific situations in your organization in which bias is currently a problem and have done the hard work to figure out clear strategies for treating everyone equitably in those situations. They may not talk about the brain areas I've just mentioned, but they don't have to. They can help you train your dACC to recognize when bias is likely to occur and they can provide your left dorsolateral prefrontal cortex new strategies for being fair so that you don't have to generate them all on your own. Because let's be honest—you already have plenty on your mind.

Unfortunately, not all unconscious bias training leads to change. One meta-analysis that examined the effectiveness of unconscious bias workshops found that training didn't always reduce biased behavior.[12] Subsequent researchers have pinpointed why unconscious bias training sometimes fails, and they found that for it to be effective, it needs to (1) help people become aware of specific situations when bias is likely to occur and (2) propose what they can do in those situations to mitigate bias.[13] In other words, it needs to target the two brain areas we just described.

Not all training includes both of those key components. So when an unconscious bias training is offered by your organization, if you're trying to decide whether it's worth your time, send a quick email asking if the training will address those two key issues: the situations in which bias is likely to occur in your organization and specific actions you can take to reduce it. You'll learn whether it's worth your time and the workshop facilitators will learn what people want and need.

3. Blur Group Boundaries

Another strategy that researchers have found to be effective is to blur the lines between your ingroups and your outgroups.[14] Essentially, you want to discover what you have in common with your outgroups. To mirror what researchers have done, get out a sheet of paper (or open up a new document on a device) and decide which outgroup you want to focus on. Perhaps you want to focus on age differences, so you pick a group that's twenty years older or younger than you. List at least five things your ingroup has in common with that outgroup. Now, in a new document, write out that list of five things again and elaborate by adding a few sentences about each one, vividly capturing what your ingroup and outgroup have in common. (Rewriting the list, rather than cutting and pasting it, helps reinforce the commonalities and blur the boundaries between the groups.) By seeing what you and members of your outgroup share, you can reduce some of the strong group boundaries that unconsciously influence you.

4. Watch a Film or Binge a Sitcom

The strategy that surprised me most was entertainment. Psychologists have found that watching a film or a sitcom with characters who are relatable, but members of your outgroup, is another effective way to reduce bias and prejudice.[15] Basically, it's a more subtle but effective way to blur boundaries. So if you're white or Hispanic, you might watch a movie or a sitcom with a predominantly Black cast. It doesn't have to be a tearjerker about racial injustice. In fact, it's probably better if it isn't. The goal here is to relate to everyday characters and see what you have in common, so it can be something light, like *Girls Trip* or *Abbott Elementary*, both funny and highly acclaimed. What's especially good news is that the effects seem to last. When white adults binge-watched six episodes of *Little Mosque on the Prairie*, a Canadian sitcom about a Muslim community, they still showed reduced bias and prejudice four to six weeks later.[16] That provides a helpful protocol: aim to watch a movie that's at least 2 hours long or six half-hour episodes of a sitcom.

One reason this strategy works, as well as the previous strategy around explicitly blurring group boundaries, is that it can help reduce how uniquely invested you are in your ingroup. You discover you have more in common with members of your outgroups than you realized, and that helps reduce ingroup favoritism. Earlier in this chapter, we saw that adults enjoy a boost of activity in their ventral striatum when they help a member of their ingroup, which means they feel more rewarded when they support someone who's like them and less rewarded when they help someone who is different. But you can stop that brain bias. Subsequent studies have shown you'll get the same level of ventral striatum activity for members of your outgroup *as long as* you see them as more like you.[17] If you feel you have more in common with members of an outgroup, your ventral striatum will light up when you help them as well. And that feels good.

***

I'm not claiming that making a list in a notebook or taking a 10-minute online test will make you bias free. I wish it were that simple. The downside to most of these self-improvement strategies is that we don't yet know if they lead to long-lasting change. Most anti-bias strategies are tested immediately, so we can be relatively confident that they reduce bias in the short term. But if you do an activity this week, will you still be treating everyone equitably three months from now, when your next big deadline pops up? It's one limitation of the research.

We're all biased, and we all have to work at reducing that bias, especially since new situations and stresses will mean you

TRY THIS
Your "Be More Fair and Less Biased" Toolkit

▶ **Take the IAT and create an action plan.** Take the Implicit Association Test (IAT) online. You'll learn which groups you're likely to show bias toward, so you'll know when you need to be more vigilant. You also want to create an equitable action plan. For example, instead of judging on a case-by-case basis which individuals receive an extension on a deadline (a situation in which you're likely to show ingroup favoritism), establish equitable criteria, such as, "Anyone who turned in their work early on the last project is eligible for an extension."

▶ **Attend unconscious bias workshops.** If your company is offering unconscious bias training, participate. It can help you identify situations in which bias is likely to occur and develop strategies to be more equitable. If you're skeptical about the training, ask the organizers if have to regularly examine whether you're treating everyone equitably or if you're inclined, however subtly, to favor members of your ingroup. It's one reason you can't just say, "Oh, I did unconscious bias training a few years ago, so I'm all set." Make time for new workshops and keep trying. By using the strategies in this chapter, you're taking steps to retrain the more conscious parts of your brain to stop the unconscious parts from running the show.

it addresses two key issues: the situations when bias is likely to occur in your organization and specific actions you can take to reduce bias.

▶ **Blur group boundaries.** Make a list of five things you have in common with one of your outgroups. Rewrite that list on a new page and add a few sentences about each item, capturing details about what you have in common.

▶ **Watch a film or binge a sitcom.** Find a movie or a sitcom about one of your outgroups and watch it for at least 2 hours. For this exercise, a show about everyday, ordinary characters is better than one about racial injustice. Your goal is to relate to the characters and see what you have in common.

THRIVE

the Rest of Your Day

Now let's turn to strategies that can help you in your personal life. We'll start by examining strategies that use brain science to help you make an unpleasant decision you've been avoiding. We'll then identify a surprising way to reduce pain as well as several tactics to reduce stress. In the final chapter, you'll learn a nonobvious approach to being a better partner when it's your significant other, not you, who's stressed.

MASTER YOUR DESTINY

CHAPTER 10

Make Better Decisions

You have a tough decision to make, one you're dreading. Maybe you need to decide whether to transition your parents into assisted living. Or perhaps you're debating whether to undergo a costly dental procedure that your dentist keeps recommending but that you've been putting off. Sure, you can research your options (for the twelfth time), but you're wondering if there's anything you can do to sharpen your decision-making toolkit.

Most of us have surprisingly few tools for making hard decisions. The one decision-making strategy most of us know is to create a trusty list of pros and cons. Sometimes a rigorous list is useful—you might uncover an important insight—but often only two or three factors truly matter to you, so the rest just muddle your thinking. You feel great about making the list, but you're no closer to deciding.

The classic pros and cons list goes at least as far back as Ben Franklin.[1] Nice, but two hundred years later? We know a lot more about the science of decision-making. In this chapter, we'll equip you with a variety of tools (including one that's quite hard to believe) that can help you make better decisions.

We're All Moody Deciders

Before we look at strategies for improving decision-making, we need to debunk a popular myth. There's a common misconception, especially in Western society, that great decisions result from thinking, not feeling. You've probably heard someone say, "Don't let your feelings cloud your judgment," as though feelings wreck otherwise good decisions.

Feelings don't "cloud" your judgment. Rather, feelings make judgment possible. The crucial role of emotions in decision-making was first made clear by Antonio Damasio, a neuroscience professor at the University of Southern California. Damasio was working with a man named Elliot who had developed a small brain tumor in his prefrontal cortex. The tumor was in an area called the ventromedial prefrontal cortex, just above Elliot's eye sockets.[2] The tumor was removed, Elliot recovered, and his IQ was roughly the same as it was before the surgery (in the top 3%).

But Elliot had two major problems. The first problem was that he no longer felt emotions. When shown pictures of severed hands or erotically posed naked women, images that evoke an emotional reaction in most people, Elliot felt nothing. He didn't just report feeling nothing; his body registered nothing. No matter how grotesque or aggressive the picture was, he showed no physiological response the way most of us automatically would. His lack of feelings made perfect sense to his doctors—when they removed the tumor, they had removed a brain region that helped regulate emotion.

But Elliot's second problem made less sense—now he struggled to make decisions. Damasio noticed this indecisiveness, so toward the end of one appointment, he asked Elliot what date and time would be best for their next appointment. Elliot got out his pocket calendar and began deliberating out loud the advantages and disadvantages of different dates, turning pages back and forth. Damasio nodded silently and watched the clock. Elliot deliberated for 30 minutes before Damasio finally cut him off. Elliot simply couldn't decide. And this indecisiveness was pervasive and crippling in Elliot's everyday life—he couldn't decide whether to use a blue or black pen when he went to write a check or how to organize the papers on his desk, decisions we hardly think of as emotional. Without any emotions or feelings to guide him, he could no longer make even the simplest of decisions.

What subsequent studies pinned down was that emotions play a surprisingly crucial role in our decision-making. Even the most analytical among us rely on the subtle thrum of emotions to make decisions, whether we realize it or not. Take professional stock market investors. It seems as though they would decide whether to buy or sell a stock based entirely on a detailed and objective analysis of a company's earning reports, new product announcements, inflation and unemployment rates, and so on. Yet people's moods actually play a huge role in these decisions. When investors are in a good mood, stock values go up. For instance, a study of stock markets in 26 countries found that when a nation had enjoyed a streak of sunny days, stock market values in that country rose. And when investors are in a bad mood, stock values go down, as evidenced by the fact that stock returns drop immediately after a country is eliminated from the World Cup in soccer.[3] Good moods and stocks go up, bad moods and stocks go down. Feelings, like it or not, shape even our most seemingly analytical choices.

Your Dentist Relies on You to Be Rational

Of course, rationally analyzing data and possible consequences also plays a central role in decision-making. If we let emotions be our only guide, there would be decisions we'd probably never ever choose. Most of us would never decide to go to the dentist, get a colonoscopy, or pay our taxes. Some of us would never visit our in-laws. Many people dread these things (or at least have more negative feelings about them than positive ones), but you know there will probably be negative consequences if you avoid these choices altogether, so you overcome your feelings and choose to do them anyway. Maybe you put these choices off as long as possible, going a full year or two (or, ahem, three) without a dentist visit, but your rational side kicks in eventually and you choose the hard road.

What Works

In a moment, we'll look at what brain science teaches us about making sharper decisions, but first, let's examine three strategies that the broader field of psychology has found can be used to improve decision-making. We don't yet know the neuroscience behind these three strategies, but we still know they work.

1. Put Yourself in the Driver's Seat

Instead of asking yourself, "What will happen if . . .?" ask yourself, "What will I do if or when . . . ?"[4] All too often, when we dread making a decision, we mentally simulate all the possible negative consequences of that decision. What's interesting here is that it's easier to picture ourselves as a passive bystander to these bad possible outcomes, someone on the side of the road watching unfortunate events unfold, than as an active adult who still has a hand on the wheel and can steer what happens next. Most big decisions are full of uncertainty, and you'll cope better with that uncertainty if you see yourself as a resilient responder, not a helpless bystander.

Of course, you can't control everything that happens. But you can improve most situations. If you decide to put your parents in assisted living, for example, what will you do if your mother feels lonely? You could help her find a club to join or agree to eat lunch there with her once a week. Psychologists find that believing in oneself makes it easier to take a risk.[5] And one easy way to believe in yourself is to put yourself in the picture as a resilient responder.

2. Give Yourself Multiple Options

Psychologists find that people are ultimately happier with their decisions when they give themselves at least two distinct options, preferably three.[6] You might think that you *do* give yourself two options all the time, but many decisions, big and small, are essentially yes/no decisions. Should I buy these shoes? Should I try a keto diet? Should this be the year we go to Paris? When you're faced with a yes/no decision, you're really only considering one distinct option (these shoes, this diet, that vacation), and whether you should or shouldn't. But when you consider two or three options simultaneously, you're likely to be happier with whatever decision you make in the long run, in part because you think more broadly about your needs. So instead of asking, "Should I buy these shoes?" ask yourself, "Should I buy these shoes, try one more shoe store, or get a foot massage with the money?"

3. Try a "Look Back"

Vividly imagine that it's twelve months from now and you're looking back on the year that's passed. Ask yourself about potential regrets or wise choices. So, you might say to yourself, "It's September 1 of next year. Looking back, I'm so glad that I . . ." Or try, "It's September 1 of next year. I would have really regretted it if we hadn't . . ." then take a few quiet minutes to see what springs to mind. You might be surprised. You might think, "I'm so glad we got my parents into a care home before the winter," or "I would have really regretted it if I hadn't made it to France." I call this a "look-back"; Daniel Kahneman, the Nobel Prize winner, calls a variation of this exercise a "pre-mortem."

It might seem counterintuitive to zoom forward and then look back, but researchers find that a wider variety of vivid details will come to mind in hindsight.[7] You already know that hindsight is 20/20, but you probably didn't know that imaginary hindsight is sharper as well. My husband and I have played this game, often around New Year's, and it's led to some great (and often surprising) choices that we're later so glad we made.

4. Slow Your Exhale, Improve Your Decisions

Now let's turn to a way to improve your decision-making that's based on neuroscience and is so simple you'll wonder why you've never heard of it. It's also so strange that you might have trouble believing it.

What's a brilliant tool for improving your decisions? Slow down your breathing.

I know what you're thinking. *I am breathing even as I read this, Therese, and this decision is still hard, so how will breathing any differently help?* It probably feels as though I've left science behind and stepped into the land of sheer speculation. What's next—should you start wearing a crystal?

No crystals, and I was skeptical of the breathing claim myself until I dug deep into the science. To understand how slowing down your breathing helps you make better choices, we need to leave the brain for a moment and talk about the vagus nerve.

Your Vagabond Nerve

The vagus nerve is part of your parasympathetic nervous system, which is often called your "rest and digest" system. As part of the parasympathetic nervous system, the vagus nerve can help calm you down by signaling to various organs that the fight-or-flight moment has passed. If you've just finished giving a stressful presentation and your heart rate is slowly returning to normal, you might literally take a big sigh of relief. That's your vagus nerve helping you relax.

"Vagus" is Latin for wandering (think "vagabond"), and it's one of the longest nerves in your body. The vagus nerve starts in the brain and runs along the carotid artery but has many branches, wandering down to the heart, lungs, and all the way to your gut. Most importantly, information flows in both directions. It's no surprise that the brain tells the heart to slow down or the lungs to breathe more slowly—you want your brain to control these organs once the stressful event has ended—but it might be less obvious that these organs also send signals back up to the brain via the vagus nerve. The lungs, heart, and stomach basically tell the brain, "everything is relaxed down here," so the brain can calm down too.

But the vagus nerve is complex. We'll say more about this complexity in a moment, but for now, it's enough to know that relaxation is just one role that the vagus nerve plays.

How Your Vagus Nerve Shapes Decision-Making

The fascinating part of all this is that the vagus nerve connects to a network of brain areas that are important for decision-making and memory. Through a chain of connections, the vagus nerve sends signals to the amygdala (a brain region that's key for processing emotions), the hippocampus (recall that's crucial for memory), the striatum and insula (areas that we've seen are important for how rewarded you feel), and to certain areas of the prefrontal cortex that are important for decision-making. Basically, your vagus nerve has a backstage pass to many regions of the brain, allowing immediate access and a lot of influence.

When your heart, lung, or gut sends a signal up through your vagus nerve that says "we're relaxing and taking it easy," that signal can improve decision-making in two ways, both directly and indirectly. First, the direct route. The vagus nerve can facilitate decision-making directly by activating regions in the prefrontal cortex that are crucial for making tough decisions.[8] Researchers find that stimulating the vagus nerve leads people to make both smarter and faster choices, even when the choices are complex.[9] In essence, the body is saying "whew, the crisis is over" and the areas of the brain that decide what to do next can now spring into action.

There's also an indirect way the vagus nerve boosts decision-making: This "rest and digest" signal also dramatically reduces stress. Stress, as you may well know, makes a mess of decision-making. As we saw in chapter 5, stress reduces cognitive flexibility, which makes it harder to change your focus from the current option you're considering to other choices. To see how stress affects decision-making and cognitive flexibility, take the simple dinner decision. Perhaps you've had an incredibly stressful week, and as you walk in the door Friday night, your significant other says, "Let's go out for dinner. Where do you want to go?" Your mind goes blank at first, then you suggest the first thing that comes to mind, a nearby Mexican restaurant you both like. Your partner points out they will have a long wait on a Friday. You're starving, but you just can't think of anything else. You're still overwhelmed from your week. You feebly suggest another Mexican restaurant, not because you're dying for a burrito but because nothing else comes to mind. Stress has put you in a decision-making slump.

And many difficult decisions in life are surrounded by stress. Deciding whether to put your parents in a care home is, in a word, stressful. Choosing to get an expensive and possibly painful dental procedure is stressful. It's no wonder you dread decisions like these and keep putting them off. It can be hard to think flexibly about your options.

Thankfully, stimulating the vagus nerve can reduce your stress and make it easier to decide. The vagus nerve helps regulate stress in the brain via multiple pathways, but perhaps the easiest one to explain is the pathway to the amygdala.[10] The amygdala is a small almond-shaped area of your brain. (You actually have two, one on the left and one on the right, but for some reason, neuroscientists call them amygdala, not amygdyli. They sit nearly level with your eyes. If you point a finger toward the most rear part of your temple, you're basically pointing at your amygdala.)

It's worth saying a little bit more about the amygdala because it's often misunderstood. You might have heard it called your "fear center," and that's because for years, researchers knew that if an animal's amygdala was damaged, scary things stopped being scary. If a monkey, say, had a damaged amygdala and you put a snake in the monkey's cage, the monkey would remain calm. But in the past two decades, scientists have discovered that the amygdala is also important for experiencing other emotions, such as feeling rewarded or pleased. A monkey with a damaged amygdala also finds it hard to learn that a grape is tasty. Snakes and grapes? Without a fully functioning amygdala, they're both "meh." Our current understanding is that the amygdala is important for figuring out whether something is a pleasure to be sought or a threat to be avoided.[11]

Let's get back to the vagus nerve and decision-making. When certain branches of the vagus nerve are activated, it sends a signal to the amygdala that there's no threat. There's nothing to fear here and the situation that was causing you so much anxiety isn't so bad after all. Your stress levels go down, and without all that anxiety driving you to consider only the safest option, it suddenly becomes a lot easier to weigh all your options and make a decision you'll ultimately be happier with.

Count Your Way to Better Decisions

Earlier I mentioned that the vagus nerve is complex. So far, I've made it sound as though the vagus nerve is your "chill out" nerve, but your vagus nerve actually has many branches, and if you stimulate certain ones, it can actually boost your energy and invigorate you. Medical researchers are beginning to use vagal nerve stimulation to arouse people from comas.[12]

I raise this point because if you do a quick search for "stimulating the vagus nerve," you'll find at least a dozen strategies. They may all have benefits, but I don't want you to chase strategies that have little impact on decision-making.

The best strategy that I've found so far, at least in terms of improving decision-making, is to slow your breathing. Since your lungs and diaphragm send signals to your brain via the vagus nerve, you want to adopt a breathing pattern that will send a strong signal of relaxation to your prefrontal cortex, your amygdala, and the rest of your brain. Several different breathing patterns have been tested, and the most effective is when your exhale is longer than your inhale.[13] Try this: Inhale deeply for 5 seconds, hold your breath for 2 seconds, then slowly exhale for 7 seconds. Some scientists call this "skewed breathing" because the exhale is longer, as illustrated below. If you're having trouble exhaling for a full 7 seconds, try exhaling through pursed lips like you're blowing through a straw, which can slow down your exhale.

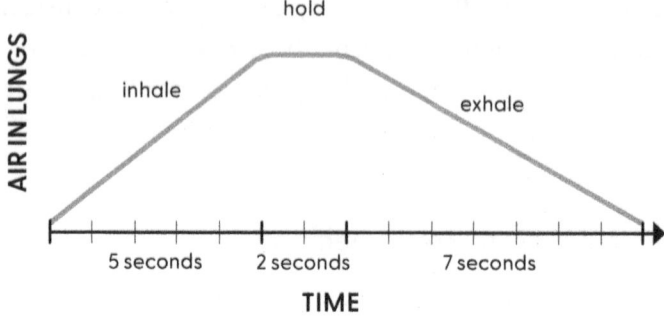

Relaxing, isn't it? Who knew you could count your way to better decisions?

If you found it hard to breathe that slowly, it might not have been immediately relaxing, but play with it. You can always start with a 4-2-5 count instead (inhale for 4 seconds, hold for 2, then exhale for 5), then work up to the 5-2-7 count.

Once you've learned how to do this kind of slow, skewed breathing, you'll have this strategy available whenever you face a tough decision. An international team of researchers found that if you do 5-2-7 breathing consecutively for 2 minutes, you'll stimulate your vagus nerve and immediately improve your decision-making.[14] In this particular study, participants were put in a stressful situation in which they had a lot of information about the staffing and organizational problems facing a fictional company, and they had to make decisions based on that information. For instance, three people had applied for the same promotion. Which person, if any, should be promoted? Although the experiment was conducted in a lab, the researchers made the scenario as realistic as possible by adding time pressure and by requiring participants to figure out which pieces of information were relevant and irrelevant. Some decisions clearly involved bad choices given the available information, but just 2 minutes of skewed breathing offered clarity and cleared the mental cobwebs. Participants in the skewed breathing conditions made a greater number of good decisions and fewer bad ones than participants in the control condition.

So take a few minutes to try the 5-2-7 breathing and come back to the decision you've been avoiding. Perhaps you'll have a new insight or at least new resolve.

Variability Is the Spice of Life

If you're a science geek like me, you'll be interested to know that scientists rarely measure the activity of the vagus nerve directly. It's possible to do, but it's complicated and invasive. What they measure instead is something called heart rate variability, also known as HRV, which usually involves putting some electrodes on your chest and connecting you to an EKG, or electrocardiogram, machine.

When it comes to your heart rate, variability is a good thing. A very good thing. Having a highly variable heart rate means that your body can adapt to the changing world around it. Your heart rate speeds up when you need to jump into action or solve a stressful problem, but it can also slow down so you can relax when you're in front of the TV, watching your favorite show. And that variability is an indirect measure of your vagus nerve's activity. High heart rate variability means the calming branches of the vagus nerve can send a strong relaxation signal some of the time—when you're enjoying that TV show, you truly put your stresses behind you. But these calming branches of the vagus nerve can also be relatively inactive at other times, allowing your body to get fired up to solve stressful problems. You don't want those calming branches to be sending a strong relaxation signal all the time because then you'd probably feel sluggish and find it hard get off the couch. You want high variability in your vagus nerve activity, and your heart rate will follow. And breathing slowly is a quick way to increase the variability of both your vagus nerve and your heart rate.

The added bonus? Practicing 5-2-7 breathing like this will improve your life in other ways as well. People with higher HRV:

- tend to live longer and have a lower risk of cardiovascular disease
- experience less anxiety
- have a lower risk of Alzheimer's disease[15]

If you're curious whether your HRV is improving with your breathing practice, you can measure it. Although an EKG provides the most accurate measurement, most of us don't have a cardiologist on call to give us those stats. Thankfully, you can also measure HRV at home using a wearable device. Chest-strap monitors tend to provide more accurate and reliable measurements than watches and are often more affordable.[16] Chest-strap monitors aren't exactly comfortable enough to wear all day long (I have worn them for running and am glad to peel them off afterward), but they are great to wear for a few hours and can

TRY THIS
Your "Make Better Decisions" Toolkit

▶ **Put yourself in the driver's seat.** Instead of asking yourself "What will happen if ...?" ask, "What will I do if or when ...?" It's easier to take risks if you see yourself as a resilient responder rather than a helpless bystander.

▶ **Give yourself two or three options.** Individuals are ultimately happier with their decisions when they consider at least two, or preferably, three, distinct options. If it's a yes/no decision, such as "Should I take this job or not?" that only counts as one option.

▶ **Try a "look back."** Imagine that it's one year from now and say to yourself, "Looking back, I'm so glad that I ..." or "Looking back, I would have really regretted it if I hadn't ..."

▶ **Slow your exhale.** Two minutes of slow breathing, in which your exhales are longer than your inhales, should immediately improve your ability to make tricky decisions. Aim for 5-2-7 breathing, in which you inhale for 5 seconds, hold your breath for 2 seconds, and exhale for 7 seconds.

give you a relatively accurate picture of whether your heart rate variability is going up over time. If you're looking for additional strategies for improving your decision-making, check out my book *How Women Decide*. In it, I examine a whole host of stereotypes about how men and women make decisions and offer strategies for rising above those stereotypes.

CHAPTER 11

Stay Healthier and Feel Less Pain

It's time for your annual physical, and you reluctantly schedule an appointment with your primary care physician. He doesn't make much eye contact and you wish he'd look at you with the same curiosity that he looks at his computer screen. Or maybe you're like me and you have a doctor who always seems a bit disorganized and a touch forgetful. Despite these minor drawbacks, you keep going back because it's such a hassle to find a new doctor. Besides, you're not exactly looking for a new bestie, so what's the harm?

The harm, it turns out, is significant.

When a doctor doesn't make eye contact or doesn't smile much, their patients have poorer health outcomes and more pain. Your doctor's bedside manner might not seem like a big deal, but if you're not pleased with it, your body won't respond as well to their treatment. A great doctor, it turns out, is one who "gets" you.

Many of us go to great lengths to stay healthy. We exercise, we watch what we eat, and we slather on sunscreen. But few of us think about switching doctors as essential to our health. However, as you'll see in this chapter, choosing a different doctor could be the best health decision you make this year.

Your Brain Can Sometimes Do the Heavy Lifting

To understand how finding a doctor you like and trust is one of the smartest things you can do for your health, we need to understand the placebo effect.

Placebo effects get a bad rap. You've no doubt heard statements along the lines of "It's *just* a placebo effect," as if to say "nothing real happened." We usually dismiss placebo effects as fake improvements, and if grandma's pain goes away with what turns out to be a sugar pill, we conclude, with an eye roll, that it was "all in her head," not in her body.

Skepticism is understandable. In the past, placebos were often used as stand-ins for real solutions. Breadcrumbs and drops of colored water were some of the many "treatments" doctors once used to satisfy patients' demands when there was no known remedy for what ailed them.[1] Given how doctors could charge high prices for what amounted to a trip to the kitchen, it's understandable why placebos gained a reputation as sham treatments.

But what doctors suspected then is what scientists are proving now: Placebos often *work*. They can alleviate pain, reduce or eliminate symptoms, and, in some instances, allow patients to be treated with half their usual dose of medication.[2] That's right—*half*.

We first need to define what a placebo effect is. Definitions abound, and many of them focus on what a placebo effect isn't. According to Dr. Fabrizio Benedetti, one of the world's leading experts on placebos, a placebo effect occurs when a patient shows a significant benefit from a treatment because brain mechanisms have been activated that anticipate improvement.[3] Something happens that makes you think you're going to feel better. This expectation activates the neural circuits involved with improving your condition, so you do feel better.

The part about brain mechanisms is important. There are many brain regions and chemicals that play a role in different placebo effects, and they're often the exact same brain regions and chemicals that make you feel better when a standard medical treatment is used.[4]

Consider pain. One of the ways that pain is reduced is through your body's natural release of opioids, which directly block the pain signals coming from your spinal cord. Placebos have been shown to increase the release of opioids, such as endorphins, in the brain.[5] There's a complex network of brain regions that trigger endogenous opioids to flood your system, but here's the abbreviated version of the story: When you expect to feel better, that expectation triggers a dopamine release in your ventral striatum. We've heard about the ventral striatum several times now. In chapter 3, we learned that dopamine is released in your ventral striatum when you're expecting enjoyment. If, for example, you pull into the parking lot of your favorite restaurant, the one with mouthwatering pizza, you'll get a dopamine release in anticipation of pleasure. Likewise, when you believe that the pill you just took should alleviate your pain, you get a dopamine release, which in turn triggers an opioid release.

This expectation doesn't just help inert placebos. This expectation can also help standard pain relievers. When you take an over-the-counter pain reliever like ibuprofen, you've essentially got two systems acting to reduce your pain—(1) there's the ibuprofen itself, which inhibits an enzyme that contributes to your pain, and (2) there are the endogenous opioids that your own body releases from your positive expectations that the ibuprofen will help. If you were convinced that the ibuprofen wouldn't help, then indeed, it probably wouldn't help as much.

So when a placebo reduces or stops your pain, it's not that you didn't have any "real" pain; it's more likely that you were able to harness your body's own system—this dopamine-opioid circuit—for dealing with "real" pain.

Hope Doesn't Cure All

There's a lot of convincing evidence out there that placebos tap into the body's innate healing systems, but for me, one finding was particularly persuasive: Individuals who show the biggest effects when taking medications also show the biggest placebo effects.[6] Some patients' bodies don't respond to certain medications at all, and fascinatingly, these patients also show no response to placebos. So it's not that hope cures all, but hope does tap into the same brain systems that *can* cure you. Still, if those brain systems aren't able to cure you, then hope won't work either.

Do placebos work for every health problem? No. An appendix that's about to burst can't be cured with a placebo and as far as we know, placebos won't lower your cholesterol or shrink a tumor.[7] Placebos are effective when the problem can be regulated by an existing brain system. But that covers a surprisingly wide range of health issues. To date, placebos have been found to reduce the symptoms of the common cold and help children on the autism spectrum communicate more easily with others. They have been used as effective treatments for everything from pain and depression to Parkinson's disease and irritable bowel syndrome.[8]

Pill or Process?

So far, it might make sense that if you believe a pill will cure your headache, that belief could help you feel better. Your belief taps into your brain's natural ability to deal with the pain. Great.

But how does that relate to your doctor's personality? As long as your doctor prescribes a standard medication, namely one you can pick up at your pharmacy, shouldn't that drug work regardless of how warm and friendly your doctor is when they're prescribing it? After all, the drug acts on specific receptors in your body, so it seems like it should be effective regardless of whether or not your doctor smiles at you.

Yet, as we hinted at in this chapter's opening, remarkably, the drug could be less effective if you don't have enough faith in your doc. You might remain sicker or in pain longer if your doctor is gruff than if your doctor is personable.

Researchers have demonstrated this for a variety of ailments. For instance, in a study of psychiatrists, half the doctors created a strong interpersonal bond with their depressed patients but treated them with a placebo, while the other half didn't try to bond with their patients but prescribed drugs for treating depression.[9] Who recovered faster? The patients with the strong interpersonal bond did, even though their medication was fake. So it's not just the pill. It's also the process, and process can trump pills. Simply put, drugs work better when you like your doctor.

I anticipate that there will be some skeptics reading this with their eyebrows raised, so I'm going to describe my very favorite experiment on this issue. It reveals the power of personality.

A team at Stanford University were doing health screenings as part of a larger study. A healthcare provider pricked everyone's skin with histamine, a known allergen.[10] Not surprisingly, participants developed an itchy raised bump and a red area around the bump. (If you've ever done a skin test for allergies, you know the drill.) Once the bump appeared, the physician rubbed in an unscented hand lotion. The lotion had no medicinal properties, but the healthcare provider told half the patients that the lotion would reduce the redness and itching and the other half that the cream would *increase* the redness and itching.

You're probably thinking that individuals who had been told the redness would subside had a smaller bump. And you'd be right. The placebo effect works for the immune system, not just the pain system. The size of the bump was significantly smaller for those who expected the cream to help them than for those who expected the cream to make it worse. For those who expected the cream to make the bump worse, it was a self-fulfilling prophecy and they developed a very itchy, angry bump. This is called the *nocebo effect,* when a negative expectation causes a negative outcome.

But that was just the beginning. What came next is where the experiment gets truly interesting.

The healthcare provider had been coached in how to treat patients and followed a specific script. For one quarter of the patients, the provider behaved with both high warmth and high competence, by doing things like smiling throughout the screening and making a lot of eye contact, using a clear, confident tone, and making no mistakes. Most of us want that doctor. For another quarter of the patients, the provider behaved with both low warmth and low competence, doing things like facing the computer for most of the appointment and not smiling, using a lot of filler words ("umm . . ."), and putting the blood pressure cuff on the wrong arm and needing to switch it. No one wants that doctor. There were also two other conditions: One quarter of the patients had a doctor with high warmth and low competence and one quarter had a doctor with low warmth and high competence.

Those who had been told that the cream was going to make the bump worse, no matter what the doctor did or said, had a bad reaction to the histamine. They had a big red bump whether the doctor was friendly or terse. So believing that they were going to react badly sent a clear signal to their bodies that even a highly competent, friendly doctor couldn't erase. It turns out that a nocebo effect can be quite strong. If you expect bad outcomes, you're more likely to get bad outcomes.

But for those who believed that the cream would help, which is what most medications and treatments are supposed to do, the doctor's behaviors *did* make a difference.

Let's first consider the people who had been told the cream would help *and* who had a highly competent, warm doctor. Their bump was a fraction of the size of everyone else's. They still had a bump—the injected histamine was going to do that—but their body was able to temper that automatic immune response, and, as a result, their skin was much less irritated. It was like a small mosquito bite, not an angry bee sting. Let's call this the lucky group.

Now let's look at what I'm calling the unlucky group. The people who had been told the cream would help but who had an incompetent, cold doctor had *terrible* reactions. They had bumps so big, red, and itchy that they looked like the group who had heard the cream would make things worse, not better. They believed the medicine would help, but their bumbling and impersonal doctor made it hard for their bodies to trust this process and harness the natural healing response. (Then again, maybe some of them doubted that the medicine would help—after all, they didn't like the doctor and the doctor was clearly making mistakes.)

What about the people who either had a competent but cold doctor or had an incompetent but warm doctor? Their bumps were smack-dab in the middle. They weren't as small as the lucky group's or as big as the unlucky group's. If you've got an impersonal or an incompetent doctor, you could be doing better in terms of your health, but you also could be doing worse.

Of course, you already know you want a competent, warm doctor because that's what everyone mentally and emotionally desires. You want to walk out of your appointments feeling like you received the care you needed. But what you perhaps didn't know was that another reason you want the competent, warm doctor is because your body will respond better to treatment. If you have a doctor you trust and you like, you will be able to summon your body's healing resources, and you won't feel as much pain or have as bad a reaction.

Incidentally, some of you might be wondering, "If hearing that a drug can make a condition worse actually makes it so, is it problematic to learn about a medication's potential negative side effects?" Indeed, this is an ongoing conversation within the medical community. It's important for healthcare providers to be transparent about a drug or treatment's potential side effects, but informing patients about those side effects does make it more likely that patients will report having them.

Oxytocin and the Benefit of the Doubt

What's clear is that it's easier to harness the body's ability to heal itself when medical professionals are both warm and proficient. The bigger question is *why*.

There are two possibilities. One is that deep down you don't really expect that an impersonal doctor can help you. If your doctor is typing away at their computer rather than looking at you, or if they seem distracted and like they're not listening when you explain the complex series of symptoms you've been having, you might think they don't really understand your problem. If they don't understand your problem, you're skeptical about their diagnosis and treatment. If you're skeptical, you don't trigger a dopamine release when you take the medication they prescribed, and you don't harness your body's natural healing pathways.

It could be as simple as that, but there's another explanation that's very promising in making sense of this phenomenon. The other possibility involves oxytocin. We can think about oxytocin in two ways—there's the very clinical side of oxytocin and there's the very social side. On the clinical side, oxytocin is crucial in childbirth and nursing. A mother's body releases oxytocin during labor to stimulate contractions, and once the baby is born, her body releases oxytocin to make milk available for breastfeeding.

But you probably know oxytocin for its social side. Oxytocin is often called the "love hormone" or "cuddle hormone" because it plays an important role in close relationships. Partners who hug each other more often have higher levels of oxytocin than partners who hug each other less, and parents who have high oxytocin levels have more positive, playful interactions with their infants than parents with low oxytocin levels.[11] And the effects of oxytocin aren't limited to human interactions. When a dog owner strokes their dog for several minutes, oxytocin levels immediately go up for both person and pet.[12]

But oxytocin isn't just important for forming bonds with the people and pets you like. It's also important for bonding with people you've just met (even the ones you don't touch). When adults were given a quick burst of oxytocin through a nasal spray, they spent more time making eye contact with a stranger than they ordinarily would.[13] Other researchers have found that a boost of oxytocin makes you more willing to cooperate with a stranger.[14] And in another, quite fascinating, study, when adults were given a burst of oxytocin nasal spray while they were playing a game with strangers, they continued to trust the strangers in the game, even when those strangers had just betrayed them.[15] Most people are slow to trust and certainly stop trusting once they've been betrayed, but the oxytocin group seemed to be thinking, "Well, I'll give this person the benefit of the doubt."

One way to understand all this is that oxytocin makes interacting with other people more intrinsically rewarding than it usually is. If you've got high oxytocin levels circulating in your brain when you meet someone, you find it rewarding to interact with them, even if there's nothing obviously rewarding about them. When your oxytocin levels are high, you find yourself thinking, "I'm enjoying this interaction," and you generously trust the other person more.

Just What the Doctor Ordered—Oxytocin

This relates back to your visit with a pleasant doctor because when a doctor is warm and friendly, you likely get a boost in oxytocin. Scientists haven't tested this directly yet, but that's one working hypothesis from neurobiologist Dan Bowling and his colleagues at Stanford University.[16] You feel a connection with your doctor, you feel cared for, and your oxytocin levels rise.

This rise in oxytocin makes you like that doctor even more, but there's another added benefit. The brain region that releases oxytocin, the pituitary gland, has direct connections to the ventral striatum, a brain area that is responsive to dopamine. So when oxytocin is released, it in turn stimulates release of dopamine. Essentially, your reassuring visit with a warm and likable doctor is triggering a chain reaction in your brain that allows you to enjoy a cascade of chemicals. First oxytocin, then dopamine, then opioids. If you've ever gone to a doctor you really trust with a problem, but then, once you're sitting in their office, find yourself thinking, "You know, this seems silly now, because it doesn't hurt as much as it did yesterday," you're not wrong. The positive experience you're having with your doctor is, through that cascade of chemicals, probably reducing your pain and your symptoms as you sit there.

That's happened to me. Once when I was traveling, I woke up with incredible abdominal pain, bad enough that I knew I needed to go to an emergency room. I didn't know the city at all, so my fiancé called around and got a personal referral for the best hospital in the area. The doctor who saw me was incredible, and I found myself telling her, "You know, the pain has lessened. Maybe I'm okay." Thankfully, she did a CT scan anyway and removed my about-to-burst appendix in emergency surgery. As I've said, the placebo effect won't cure appendicitis, but it can reduce the pain associated with it.

What Works

The long and the short of the preceding discussion is that if you aren't terribly fond of your doctor, find a new one. You may not be having many health problems now, but once you have a doctor you like and respect, you'll have a faster road to recovery if and when you do get sick.

What to Look For

So what should you look for in a medical professional if you want to tap into your body's own healing abilities? A warm, personable doctor is key. In the research lab, healthcare providers show high warmth by doing the following:[17]

- Calling the person by name and introducing themselves in a friendly way
- Making frequent eye contact
- Sitting close to the patient
- Only looking at their computer occasionally
- Smiling
- Having soothing or uplifting art and images on the walls.

Recall that credibility also played a role in the body's ability to heal itself. In the lab, researchers increase credibility by having the healthcare provider do the following:

- Using a clear, confident tone when speaking
- Not using filler words (such as "um")
- Making no mistakes during the appointment
- Having an organized, neat, and clean room

Smiling—*your* smiling this time—is another good sign. Researchers have found that smiling improves your health,[18] so pay attention to whether you find yourself smiling easily during an appointment (and whether you're smiling because you truly feel happy or because smiling is what you feel you're supposed to do).

Keep in mind that you have your own personal preferences and pet peeves. Maybe you relax when a doctor asks you about your family, or maybe you think that's weird. Maybe you respect a doctor who has diplomas on the wall and who talks about their accomplishments, or maybe that feels like boasting. Ultimately, you want to feel good about the interaction, so notice whether you leave a doctor's visit feeling glad that you came or just glad that it's over.

Do you need to love every doctor you see? That's not realistic. If you have a suspicious mole and see a dermatologist twice, once for the initial consult and a second time to have it removed, you don't need to bond with that doctor for that procedure be effective. Or if you need to have a procedure urgently, then there's no time to shop around (though finding a highly reputable hospital, as my fiancé did for me, might be worth an extra 10 minutes). But if you're going to be seeing a doctor regularly, or you don't need immediate care for an acute condition, you'll have better outcomes if you take the time to find someone you like.

**

I know that finding a new doctor is a hassle. When I first started writing this chapter, although I was convinced I needed a new primary care physician, I could feel myself making excuses to not go out and find one. But after reading the research, I tried two different PCPs before finding one I really like. Now I feel heard, seen, and cared for. Finding a doctor you like and trust might be a pain in the neck, but you're saving yourself from real pain down the road.

TRY THIS
Your "Stay Healthier and Feel Less Pain" Toolkit

▶ **Find a doctor you like and trust.** You want to tap into brain systems that reduce pain and support healing, and one powerfully surprising way to do that is to find a doctor that you like and trust. If you don't enjoy their personality or you sometimes doubt their competence, their treatments will be less effective.

RELAX AND CONNECT

CHAPTER 12

Take Charge of Chronic Stress

Money is tight. Your grocery bill keeps rising, your car needs some long-overdue repairs, and your day care just announced a rate hike. You don't even want to think about the bathroom and the dripping faucet. You're in no position right now to look for a job that might pay more, and you increasingly find yourself lying awake at night worrying how you'll pay for it all (and closing the bathroom door so you can't hear that drip).

Life is stressful. Whether it's financial struggles, pressures at work, health concerns, relationship issues, or upsetting events in the news, we all have times when the demands of life exceed our ability to cope and stress can be crushing. If there's one area in which most of us could use sharper strategies, I'd put my money on stress management. In the next two chapters, we'll examine how to manage stress so that you can enjoy your life more.

In this chapter, we'll consider how to cope with chronic stress, like money woes, that can last for weeks, months, or in the worst case, years. In the chapter that follows, we'll look at acute stressors, like a visit to your in-laws or a presentation you have to give, experiences that may be upsetting but are typically resolved in a few minutes, hours, or days.

WHAT DOESN'T WORK

There are plenty of unhealthy ways to manage your chronic financial stress, everything from letting unopened bills pile up to drinking an entire bottle of wine at the end of the day. Psychologists call these "avoidance strategies" because they're strategies to avoid whatever is stressing you. Avoidance strategies like these may improve your mood in the short term, but as you might know from watching a friend or loved one, if an unhealthy avoidance strategy is used too frequently, it can amplify the old stressor (which, of course, hasn't gone away) *and* create entirely new ones.

You already know that drinking through your problem isn't a good way to go, but you might be wondering about other avoidance strategies. For instance, is distracting yourself from your stress by doing one of your favorite hobbies or by browsing the internet over lunch a healthy way to cope? After all, these distractions make you feel better, at least short-term, and aren't as self-destructive as, say, eating an entire pint of ice cream.

It turns out that distraction is a delicate balance.

New research shows that it depends on why you're turning to that distraction.[1] If you're turning to it to increase your positive mood, then the distraction tends to be a healthy coping strategy. If you spend Saturday morning baking cookies in your kitchen or shooting arrows at the archery range because it lifts your spirits, calms your mind, and makes it easier to tackle some of your money problems on Saturday afternoon, then keep that up. The positive mood can refill your mental and emotional tank and give you the energy to reengage with the stressful situation later.

But if you're turning to that distraction to avoid your money problems and the stressful situation, then the distraction tends to be an unhealthy coping strategy. You're likely to feel more depressed and stressed, not less, when you take your last tray of cookies out of the oven or shoot your last arrow for the day because your escape has ended. If you find yourself thinking, "I really wish this problem would just go away and I don't want to think about it," then time spent on your hobby is probably increasing how stressed you feel, not decreasing it.

Your motto when it comes to distractions should be, "I'm doing this to recharge, not to escape."

It's tricky, of course, to honestly assess our own motivations. Most of us can rationalize just about anything. To gauge how inclined you are to use avoidance strategies, read the following sentences and give each sentence a score of 1 to 4, with 1 being "I usually don't do this" and 4 being "I usually do this a lot."[2]

___ I wish the situation would go away or somehow be over with.

___ I avoid being with people in general.

___ I make light of the situation or refuse to get too serious about it.

___ I turn to work or other activities to take my mind off things.

___ I admit to myself that I can't deal with it and quit trying.

Your score could be anywhere from 5 to 20, and the higher your score, the more you're motivated by avoidance and the more likely it is that your distractions are increasing your stress instead of mitigating it. If you have a high score, the strategies later in this chapter may be hard to adopt, but they could also be a lifesaver.

On a Hamster Wheel, Literally

To understand how to manage stress in healthy, productive ways, we need to talk about hamster wheels. Perhaps you've heard the phrase "hamster wheel" used metaphorically to describe repetitive, mind-numbing tasks, as in "I feel like I'm trapped on a hamster wheel at work."

But we're going to look at real hamster wheels, the kind that rodents run on, because these little whirring wheels actually teach us something incredibly enlightening about how to *avoid* stress. Researchers often place mice in hamster wheels (or on tiny treadmills) to see how exercise, particularly running, affects their health. Before these studies were done, the general thinking was that mice that exercised would generally be healthier than mice that didn't. Runners are a healthy lot, right? Researchers expected that there would be a variety of cardiovascular and muscular benefits from running, plus, as we saw in chapters 5 and 6, moderate to intense aerobic exercise benefits your brain.

They quickly discovered, however, that not all running is created equal. When mice were put in a cage with a hamster wheel and had the option to run when they wanted to, those mice reaped the benefits you'd expect, plus a few more. They had healthier hearts and held onto more muscle as they aged.[3] They also showed fewer symptoms of depression. The depression findings are especially interesting: When mice were put into unpredictably stressful conditions, such as being deprived of food and water for 24 hours, or when their cage was tilted at a 45 degree angle for hours on end, the stressed-out mice who had periodic access to a running wheel coped much better with their stressful living conditions than the mice who just had to endure the awfulness.[4] When given the option to run, they had fewer depression-like symptoms compared to mice with no wheel.

You might think that if mice had gone 24 hours without food or water, they would be too physically exhausted to hop on the wheel, especially since mice have much higher metabolisms than humans. Their metabolisms are about 7 times faster than ours, so 24 hours is a very long time without nourishment for a mouse.[5] But hop on the wheel they did. The workout reduced their stress. (If you've ever had a ridiculously stressful and exhausting workday but then stopped at the gym for 30 minutes and walked out feeling worlds better, you understand.)

Those were the mice that had the option to run. But what about the mice that were forced to run? They had a very different experience, a largely unhealthy one. When mice were put on tiny moving treadmills and had no choice but to run, their existing health problems became much worse. If a mouse had colon problems and was forced to run, the colon problems became aggravated. Interestingly, the colon problems would begin to go away if a mouse could run when it wanted to.[6] If a mouse had Alzheimer's-like plaques and was forced to run, the plaques got worse; but again, the plaques began to disappear if the mouse could run when it chose to.[7]

It might seem like the obvious conclusion here is that you should only exercise when you want to, and that's one possible takeaway. But there's an even bigger takeaway: The absence of choice and the absence of control are incredibly stressful. When one feels forced to do something, an activity that's ordinarily stress-relieving becomes stress-inducing. In the case of the mice, instead of benefiting their bodies, the exercise they were forced to do led to an increase in stress hormones, causing extensive tissue damage.[8]

But when one chooses to engage in an activity, even an activity that is very hard, that choice and control can have a healing impact. If you want to manage stress, you don't want to be a mouse that's forced to run. You want to be a mouse that chooses to run. You want control, and you want choice.

Why People Like to Pick Their Lottery Numbers

You may be unimpressed by research with mice, so let's turn to research with humans. Not surprisingly, most of us like to control our environment and our fate. If you're cold, you want to be able to turn up the heat. If you don't like what's on TV, you want to be able to change the channel or at least watch something else on your phone. You don't want to be told you just have to sit there and endure a situation you dislike (as any teenager will tell you).

The benefit of having choice and control isn't just that you like your environment a little more. Choice is immediately rewarding in the brain. People like choice and control so much that when they're anticipating a choice, even before they've made it, reward centers of the brain light up.[9]

That last sentence is fascinating. We'd expect that if you made a choice and you received something delightful, say £1,000, because you chose wisely, then the reward centers of your brain would light up. It's obvious why £1,000 lights up those reward centers—you're taking a chunk out of your money problems! But what's less obvious is that simply looking forward to choosing also lights up these reward centers, even when there's only a small chance you'll get that £1,000. Successful choices you've just made feel good, but so does the possibility of future choices you haven't made yet. Our brains are designed to seek out choice and control.

For all of us who don't understand why people get excited to pick their lottery numbers, now it might make a little more sense. The chances of winning are incredibly low, but you're choosing your numbers. And choosing itself is rewarding.

Choice isn't just immediately rewarding for your brain. Just as we saw in mice, choice and control have long-term benefits for your body as well. When people have more control over their lives, they live longer.[10] They have healthier hearts, fewer doctor's visits, and less drug addiction. People who have a lot of choice are mentally and physically healthier than those who don't.[11]

Perceived Control
(Because Life's Not All Cat Food)

Does that mean that the healthiest people are billionaires, those who have more choice and control than the rest of us? Not necessarily. It depends on how they see their situation. A billionaire may feel trapped in maintaining their wealth and status in the public eye, which might make them feel like they have less control and fewer choices than someone who has less money. Don't worry, I'm not trying to get you to feel sorry for billionaires, I'm just noting that choice, like beauty, is often in the eye of the beholder. What social scientists have discovered is that *perceived* control is much more important to your health than *actual* control.

Perceived control is a belief that you can influence your environment, your relationships, the events in your life, your behavior, and your own internal states.[12] Someone with high perceived control believes they have choices and if they make the right choices, they can improve their situation. Let's say you're a student. If you perceive you have very little control, you may believe that no matter how hard you work and no matter what you learn, the system is stacked against you, and you can't earn a grade better than a C+. But if you perceive you have a lot of control, you might believe that by working hard and mastering the course material, you can earn good grades.

That's perceived control. Actual control, in contrast, would occur in parts of your life where you have the power to completely determine the outcome. If you own a cat, you have actual control over whether or not you feed your cat. Barring some odd and unlikely event (like someone recklessly throwing out all the cat food), you do have control over your cat's meals. You don't, unfortunately, have *actual* control over your school grade because the professor may not have been clear about the grading criteria or other students may have relevant expertise you lack, either of which could lower your grade.

As you might imagine, most of us have little actual control over many events in our lives. Life isn't all cat food. You can leave work on time, which might be under your control, but whether you arrive home on time depends on a dozen things, from traffic to road construction to whether your best friend texts you in a crisis.

But no matter our circumstances, we can have a lot of *perceived* control. Even a cancer patient can perceive that they have a high degree of control. They may not be able to control their diagnosis, but once diagnosed, they can perceive control over their relationships with friends and family, their treatment plan, and their access to medical information. They can decide, for example, who in their family they want to tell about their symptoms, even though they can't control what those family members share with one another. But by increasing their perceived control, perhaps by saying "Please keep this between us," they have more control over their experience of the disease.

Or they can perceive that they have little control over all of this, that whatever happens is outside their influence. If you have low perceived control, you feel like your fate is not in your hands.

You want to be in the first group. Research reveals that cancer patients who perceive they have a high degree of control are a lot less depressed and distressed than those who perceive they have little control. And cancer patients with low perceived control feel much sicker, regardless of the severity of their physical symptoms. When patients with low perceived control have relatively mild physical symptoms, they nonetheless report feeling worse than patients with high perceived control who have more severe symptoms.[13] Perceived control has a bigger impact on how patients feel than their physical states.

Perceived control is protective, like your little shield against life's onslaughts.

Feeling in Control Makes the Difference Between Hopeful and Hopeless

Perhaps you don't have cancer, but you do have money problems, and that's stressful enough. The good news is that your stress levels are likely to go down dramatically if you can increase your perception of control. Psychologists who study stress find that effective coping strategies, from exercise to therapy, typically increase an individual's sense of control.[14] Your tone needs to become one of "taking charge." In a moment, we'll look at techniques to dial up your perceived control.

But first, let's understand the brain areas that help you feel less stressed when you feel a sense of control. The first important brain area is the ventral striatum. As you might remember from chapter 3 on motivation, one role of the ventral striatum is to anticipate that something will be pleasing or gratifying in the future. The ventral striatum is one of your reward centers. And knowing you get to choose soon? That's rewarding. As we've already seen, simply feeling that you have a choice feels gratifying all by itself, even before you've chosen a darn thing. And neuroscientists find that even believing you will get to choose soon lights up your ventral striatum and makes you feel hopeful.[15]

Putting people in a situation in which they have some choice, however, does not light up everyone's ventral striatum equally. Individuals who see the choice as an opportunity to take one or more possible paths, and perceive they have a lot of control in picking which path to try, will have a lot of activity in their ventral striatum.[16] People who have those exact same paths in front of them but don't believe they can take them will feel helpless and have very little activity in their ventral striatum. For instance, if someone is having financial problems, a friend might suggest calling the bank to discuss their options. But if that person can't imagine calling up their bank to have that conversation—perhaps they are too embarrassed—they will feel very little control. If the choice in front of them doesn't seem feasible or within reach, they feel very stressed.

The key takeaway here is that simply seeing choices out there in the world isn't enough to light up your reward centers.

You have to believe that those choices are realistic for you and that you have some control to go after them.

What Works

How do you increase your perceived control? Psychologists often discuss what's known as "locus of control," or the general degree to which a person perceives they can control the events in their life.[17] At one end of the continuum are individuals with a strong *external* locus of control, individuals who believe the external world sets their fate. Chances are you can think of someone in your life with a strong external locus of control, someone who feels they're at the mercy of the world and can't do much to influence their circumstances—the kind of person who sits in a restaurant complaining because there's a lipstick stain on their water glass, but won't do anything to grab the waiter's attention. At the other end of the continuum are individuals with a strong *internal* locus of control, who believe they determine whatever comes their way. Someone with a strong internal locus of control might go to great lengths to get a waiter's attention, waving vigorously or walking up to the waiter with the glass in hand.

You're somewhere along that continuum. Even though you probably have a default way of behaving, researchers are finding that whatever your default is, there are some ways to increase your perceived control when you find it lagging.

1. More Gain, Less Pain

The first strategy I'm going to suggest isn't obvious, but it's powerful. An important way to increase your perception of control is to reframe your situation in terms of what you have to gain. Researchers at Harvard and Rutgers University find that when a choice is framed as "what could I gain?" individuals feel like they have a lot more control over the situation than when the very same choice is framed as "what could I lose?"[18]

An example here will help. Imagine that you realize you won't be able to pay all your bills this month. After you pay your rent, you could pay your two big bills or your three smaller bills, but you can't pay all five. You'll have to choose what to pay and what not to pay. The question is, how do you frame that choice in your mind?

If you frame that choice as "no matter what I choose, I can't pay all my bills," then you're framing the situation as a loss. You've told yourself you're in the hole no matter what you do. With this framing, you're not going to feel you have much control over your money problems, and you're going to feel stressed and potentially helpless. Your ventral striatum is unlikely to light up—it isn't interested in losses—so you won't feel rewarded, no matter what you choose.[19]

But if you frame that same choice in terms of what you could *gain,* then you'll feel you have more control. You could frame it as, "I'm going to make the choice that will leave me with £25 to go out for pizza with my friends on Friday." Simply by focusing on what you'll have to gain, you'll feel you have more control, and your ventral striatum is much more likely to light up, making you feel rewarded.

Either way, you're paying the same bills now and letting the same bills go until next month, but in one case your stress levels go down, and in the other case your stress levels stay high. Reframing in terms of what you could gain won't fix your financial problems, but it will likely put you in a better place to resolve them.

DOES AGE MATTER?

Life would be lovely if your sense of perceived control kept increasing with each passing decade, but anyone over seventy will tell you that's often not the case. Perceived control does generally increase in a person's twenties, thirties, and forties (look at you, building your career, hobbies, and family!) which can be delightful. It tends to peak around midlife, begins to drop slightly around retirement age, and then, unfortunately, drops steeply in a person's seventies and eighties.[20]

Many factors contribute to that loss of perceived control. Retirement means you don't have status at work anymore. Back pain and arthritis make everything harder. Technology keeps changing and you're proud to have a Snapchat account, but why can't you turn on your TV?

Research documents that if you can maintain a relatively high level of perceived control despite all of these challenges, you'll benefit in spades. You'll have better mental and physical health after age sixty-five, you'll live longer, and your memory will stay sharper.[21]

Clearly, some older adults have higher levels of perceived control than others. What's their secret?

There seem to be three. First, our tip about focusing on gains over losses seems to be key, especially as we age. Older adults who focus on gaining control over the outcomes they *do* want enjoy a much greater sense of control than those who focus on gaining control over outcomes they don't want.[22] Let's unpack an example. You want to become more physically active, but you have chronic back pain, so your doctor gave you a list of rehab exercises. The key question is this: What reason do you give yourself for doing those exercises? You could focus on how you want to avoid pain flare-ups, which would be gaining control over something you don't want. Or you could focus on your growing ability to do longer bike rides every weekend, which would be gaining control over something you *do* want. (Plus, you watch your favorite streaming show when you do your rehab every afternoon, which is an added bonus.) If you embrace those latter motivations, because you're gaining something, you'll have more perceived control. Now, why do you do the rehab?

Our second and third tips, below, both relate to social isolation. Older adults who are lonely have less perceived control. Feeling like you have no friends leaves you adrift. So if you're looking to do more physical activity, pick something social, like cycling with a group or, as we saw in chapter 6, taking a dance class. Finally, older adults who do more volunteer work have higher levels of perceived control.[23] Volunteering might boost perceived control in part because it keeps loneliness at bay, but volunteering does more than that. It also gives you a chance to learn new skills and use old skills that are otherwise sitting dormant, both of which can help you see that you have influence over your life.

---

2. Name It to Tame It

Let's return to cancer patients to discuss a second strategy on how to increase your perceptions of control. Cancer patients have low absolute control because some cancers are degenerative and fatal. But as we saw earlier, even cancer patients can have high perceptions of control. Some feel helpless, but those who don't feel helpless cope much better with their disease.

Research with cancer patients reveals that the biggest benefit comes from seeing their emotions as under their control.[24] And that's something you can do. You may not be able to do as much as you'd like to resolve your money problems, but you can control your emotions about your money problems. If you're feeling anxious, what strategies do you have for dealing with that anxiety? Mindfulness meditation, discussed in chapter 6, helps many people reduce anxiety.[25] The breathing techniques discussed in chapter 10 have also been shown to help. If you're feeling angry about your money situation, do you have strategies for dealing with your anger? Exercise can be an extremely effective way to manage anger without repressing it.[26]

If you're feeling chronically stressed, you might be convinced that reducing your distressing feelings is going to require time and energy that you simply don't have. You may be thinking, "You expect me to learn how to meditate or drive to the gym with all that's going on?"

I have good news. One of my favorite ways to control troubling emotions is simple and quick: *Name* that emotion. A growing body of research reveals that reflecting on and naming your distressing emotion is a sneaky but effective way to tame it and bring it under your control.[27] "You've got to name it to tame it," as some psychologists like to say.

The key here is to get nuanced and specific. Simply saying "I'm upset" or "I feel terrible" is too vague and unlikely to help. Figuring out the more nuanced emotion you're feeling can go a long way to reducing its intensity and helping you manage it. When you think about whatever is stressing you right now, are you feeling more worried, embarrassed, frustrated, or panicked? Or maybe another word will describe it better.

For naming emotions, I love the How We Feel app. Founded by a science-based nonprofit out of Yale University, it prompts you to check in about how you're feeling so that you become more aware of your feelings throughout the day and helps you track which activities and environments improve or tank your mood. Plus it's free. I started using the How We Feel app in early 2024, and it's one of a handful of tools that I use almost daily. Like many women, I struggle when I'm feeling angry, and learning to name it has helped me respond to anger more productively. I especially love its brief video lessons on coping with troubling feelings.

3. Increase Activity in Your "I Think I Can" Brain Area

The strategies above are related to increasing your sense of perceived control in order to activate your ventral striatum. There's a second key brain area related to stress and control that's worth knowing about, a region called the ventromedial prefrontal cortex. I know—it's a mouthful. Neuroscientists usually abbreviate it as the vmPFC. (I like to jokingly think of it as your "runDMC" brain area.) Perhaps you remember Elliot, the man from chapter 10 who found it so hard to make decisions that he couldn't even schedule his next doctor's appointment and would debate his options for 30 minutes and still not land. He had brain damage to an area above his eye sockets, in his vmPFC.

The vmPFC helps you gauge an answer to the question, "Given how I'm feeling, what do I want to do?" And when you're stressed, you really have two broad options: (1) you can try to do something or (2) you can give up. For instance, you can keep trying to do something about your money problems or you can say, "There's nothing else I can do." Neuroscientists find that if you're stressed but your vmPFC is firing on all cylinders, you keep trying.[28] Even if you don't have a lot of actual control over the stressful situation, if your vmPFC is firing, you persist and try something you haven't tried before, such as asking ChatGPT, "What are 10 creative ways introverts can deal with money problems?" It's as though the vmPFC is fueling the little engine that could. It's taking how you're feeling, which is probably pretty crummy, and converting those negative emotions into "I think I can. Or, at least, I've got to try."

But when your vmPFC goes quiet, you give up. You don't die. But you don't try. When individuals have little activity in their vmPFC, they give up relatively quickly and show no interest in trying new strategies.

Here's especially promising news, however: If your vmPFC is highly active when you're stressed, you'll rebound more quickly. People who have a highly active vmPFC during a stressful situation recover from that stress sooner than people who have a quiet vmPFC. Let's say you're on vacation in a beautiful place, perhaps an island you'd been dreaming about, and you open your wallet and discover a credit card is missing. It's highly stressful. You call your bank and get the card canceled and a new one ordered, but once the peak stressful moment has passed, do you rebound quickly to a good mood and go back to enjoying your trip? Or do you ruminate, upset for hours and unable to let it go?

Neuroscientists have discovered that the vmPFC plays a key role in determining how you react. If your vmPFC is highly active when you're in the middle of a stressful situation, you'll experience more positive emotions and fewer negative emotions once that stressful situation is over.[29] You reframe or reappraise the situation, perhaps thinking, *that was stressful but it could have been much worse,* and put it behind you. But if your vmPFC is less active during that stressful situation, you are more likely to stay anxious and find it hard to enjoy your precious vacation. So, you want a highly active vmPFC during stressful situations because it will fuel you to keep trying and help you return to a positive mood once the peak stressful situation has passed.

From all this, you're no doubt thinking, Well, how do I increase the activity of my vmPFC? It sounds like a one-stop shop for resilience and for persisting when the going gets tough. The two stress-reduction strategies below show you how.

Slow Breath Out. One strategy for increasing activity in the vmPFC is to increase your heart rate variability. We learned about heart rate variability (HRV) in detail in chapter 10, on decision-making, so I won't rehash all of that here, but recall that one way to increase your heart rate variability is by doing some skewed breathing, such as 5-2-7 breathing. Inhale slowly for 5 seconds, hold your breath for 2 seconds, then exhale slowly for 7 seconds. If you do this for 2 minutes, your heart rate variability temporarily goes up.

Why do you want to your heart rate variability to go up? What we learned in chapter 10 is that when your heart rate variability increases, your decision-making improves, which is certainly helpful if you're facing some stressful situations. But there's an added benefit. A different team of researchers have found that individuals with higher heart rate variability also have greater activity in their vmPFC.[30] The exact mechanism isn't understood yet, though chances are it's related to the vagus nerve.

So when you're stressed out, don't just take a deep breath, take a 5-2-7 breath. If you can do 5-2-7 breathing for 2 full

---

TRY THIS
Your "Take Charge of Chronic Stress" Toolkit

▶ **Ask yourself, "What could I gain?"** When you're considering a possible choice you could make or action you could take, instead of focusing on "What could I lose?" ask yourself, "What could I gain?"

▶ **Name it to tame it.** See your emotions as under your control. Practice being more specific and nuanced in naming your feelings as they arise. An app like How We Feel can help.

▶ **Increase activity in your "I think I can" brain area.** Increasing activity in your vmPFC should help you feel more resilient and be willing to persevere in trying to resolve your stressful situation despite setbacks.

> **Slow your exhale.** Improve your heart rate variability by practicing 2 minutes of 5-2-7

minutes, you're more likely to stimulate the brain area that gives you resilience and perseverance.

Go to Your Happy Place (Temporarily). The second strategy for increasing activity in your vmPFC may seem like a clunky version of a pop song: If you want to be resilient, don't worry, be happy. Neuroscientists find that taking a few minutes to get in a happier or more excited mood increases activity in the vmPFC.[31]

---

breathing (inhale for 5 seconds, hold for 2, exhale for 7).

> **Improve your mood.** Do something that makes you happy, such as watching a funny video or engaging in a hobby; then, while you're still feeling happy, return to working on the situation that's causing your stress.

▶ **Does age matter?** Older adults see boosts in perceived control when they focus on gaining control over an outcome they want, rather than on gaining control over an outcome they don't want. Volunteering and reducing loneliness also increase perceived control in people over age sixty-five.

(A positive mood also increases activity in the ventral striatum, so it's a twofer.) In the lab, scientists boost moods and activity in the vmPFC by having participants listen to upbeat music and by reading a long list of positive sentences (e.g., "I feel amazing today!"). You could lift your mood by listening to some favorite music or distracting yourself with a hobby or a funny video.

The temptation, however, will be to just keep distracting yourself, riding this good mood for as long as it lasts, and that's not why you're doing it. The trick is to go back to dealing with your money problem or whatever is stressing you while you're still in that good mood and your vmPFC is firing away. The goal here is to tap into this brain area's ability to help you persist and find a more permanent solution to what's stressing you.

I realize this might feel a bit like a chicken-and-egg problem. Earlier I said if you can increase activity in the vmPFC, then you'll enjoy a better mood. So which comes first? Does increased activity in the vmPFC improve your mood, or does a good mood activate the vmPFC? Fair question. There seems to be evidence for both possibilities. Hopefully future research can tease apart the mechanism with greater precision, but for now, if you can find something that makes you happy, you should feel more resilient when you go back to tackling what's stressing you.

Neuroscience might not be perfect, but it can still be incredibly helpful.

CHAPTER 13

Manage Intense Moments of Acute Stress

Let's say you're home for the holidays, seated around a festive, crowded dinner table, and your mother-in-law says something that really upsets you. Your heart beats faster and your jaw tightens when she starts attacking your politics, criticizing your career, or judging your parenting. Your spouse would normally intervene, but he's deep in a side conversation. You're tempted to make an excuse and leave the table because you're worried you'll launch a nasty rebuttal if you stay, but you'd really like another way to cope with stressful moments like this other than simply hitting eject.

We've all heard of the fight-or-flight response, and in this situation, your body seems to be preparing for one or the other. Thankfully, neuroscientists are discovering that our coping strategies go far beyond f-words.

The strategies in this chapter are all about handling those acutely stressful moments that arise in our personal lives—from having an argument with a family member to anticipating an anxiety-provoking medical procedure. But as you'll see, some of these strategies can be easily applied to work stresses too.

What Works

Touch, it turns out, works wonders. The first three strategies for reducing acute stress all draw upon the healing power of touch, and the last three strategies take other approaches.

1. Hug for Health

One of the simplest stress relievers you can do before walking into a potentially stressful situation is to hug your partner. Not a quick side hug while you're talking with someone else or a perfunctory pat on the back, but a good long embrace in which you focus on each other. Simple, yes, but research shows it's also surprisingly effective—at least it is for women. (More on men in a minute.)

Although many studies have revealed that gentle, support-ive touch reduces stress, let's look at one of the simplest studies conducted. An international team of researchers brought cou-ples who were happy in their relationships into the lab.[1] Half the couples were asked to embrace for 20 seconds in a waiting room and half were asked to wait for 20 seconds in that same room with no additional instructions. (The latter group could spend the time however they wanted, but if you've ever sat in a waiting room with your partner, you know that an extended hug probably isn't your go-to.)

Then they were all put in a stressful situation: Each person had to submerge one hand in ice-cold water (32–39 degrees Fahrenheit) and keep it there for several minutes. Ice water submersion is a quick and surefire way to make an individual highly stressed. It may not sound as stressful as being grilled by your in-laws, but trust me, one's blood pressure and cortisol levels skyrocket.

We haven't discussed cortisol yet, but if you've done any reading about stress, you've heard of it. Cortisol is your body's main stress hormone—it helps your body stay on high alert when you feel threatened—and the amount of cortisol in your bloodstream is an objective indicator of your stress levels. Although many factors can affect your cortisol levels, generally speaking, the more stressful events you experience over the course of a day, and the more distressed you feel about those events, the higher your cortisol levels.[2]

Hugging, however, helps. The women who had been hugged before putting their hands in the ice water had significantly lower cortisol levels when they pulled their hands out than the women who had just sat and waited with their partners. Hugging seemed to act as a protective buffer, a partial suit of armor if you will, soothing the women's nervous systems and making them feel less threatened by the stressful event.

Surprisingly, the hug didn't lower men's cortisol levels. I say "surprisingly" because other researchers have found that a 10–20 second embrace *can* be enough to lower blood pressure, heart rates, and cortisol levels in men, but hugs more consistently reduce stress for women.[3] Women tend to rate hugs as much more pleasing than men do, and that may partly explain why women's bodies react differently.[4] Men who enjoy a good hug probably show the same stress-relieving effects that women do.

Mind you, all participants still found the cold plunge unpleasant. Everyone's mood soured a little, hugged or unhugged, so it's not that it gave anyone a complete suit of armor. But hugging a partner whose touch you enjoy before you walk into your in-laws could make stressful confrontations a little easier to take.

Some readers will think, "Great strategy, if only I had a partner to hug." Maybe you have a partner who would never agree to a 20-second hug, or maybe you don't have a partner at all. Good news—the hug doesn't have to come from a partner. One research team has found that even a 20-second hug from a stranger is enough to lower cortisol levels surrounding a stressful event.[5] Most of us can't find a stranger we'd want to hug for 20 seconds, but you might be able to ask a good friend.

2. Soothing Self-Touch Works Too

These same researchers found that soothing self-touch was just as effective at lowering cortisol as an extended hug. Touch can be soothing to your nerves, even when it comes from your own hands. In this study, most participants chose to put one hand on their heart and one hand on their stomach, but soothing self-touch might involve hugging oneself and gently rubbing the tops of one's arms or placing one's hands gently on one's cheeks and forehead. Experiment a bit and find the kind of touch that makes you feel calm, comfortable, and relaxed. Once you find something that feels right, keep your hands there, possibly moving them in a slight caress, for 20 seconds, and concentrate on the warmth of your hands and rhythm of your breathing. You'll probably feel calmer and steadier, and whatever stressful event follows, it shouldn't be nearly as triggering.

3. Stock Up on Affection

There is one important caveat about these studies using touch: timing. In the studies I've just described, people soothed themselves with hugs or self-touch immediately before the stressful event. They lingered in that hug or self-touch for a full 20 seconds and a few minutes later, they were put in the stressful situation. A protracted 20-second self-hug in the dining room right before you sit down for dinner might be weird (although you could sneak off to the bathroom and wrap your arms around yourself in private).

The good news, however, is that researchers find that the more affection you receive over the course of the day, the lower your cortisol levels tend to be. Specifically, people who reported receiving more affectionate communications from their spouse, either in the form of affectionate touch or verbal reassurances such as "I love you," had lower overall cortisol levels than people who reported receiving less affection.[6] So try stocking up on hugs, kisses, and loving reassurances either the day before or the morning of a potentially stressful event. With lower cortisol levels going into the conversation, you'll probably feel less threatened by unwelcome comments.

4. Every Breath You Take

Let's turn to some ways to reduce your stress that don't involve touch or affection. One effective and easy strategy is what's known as voluntary or controlled breathing. You're already breathing involuntarily (look at you go!), and the idea here is to adjust your breathing to tap into your body's ability to calm itself.

Scientists find that slowing down your breathing can dramatically reduce your stress levels.[7] In chapter 12, we saw how 2 minutes of slow skewed breathing is one way to activate your ventromedial prefrontal cortex and give you resilience when you're facing chronic stress. Now we're looking at acute stress, and although skewed breathing won't hurt, the best breathing techniques for acute stress differ in two ways.

Let's say that you know you're likely to face a stressful situation. (After all, this isn't the only time your in-laws have grilled you.) First, you want to do slow breathing for a little longer than 2 minutes. For acute stress, scientists find that 5 to 10 minutes is much more effective.[8] In one study conducted at the University of Pennsylvania, young adults who were about to walk into a tense situation were instructed to do very slow, deep breathing for 5 minutes. They took a break for a few minutes, then practiced slow breathing for another 5 minutes, then faced a stressful situation. The deep breathing practice beforehand helped the adults maintain a lower heart rate despite the stress and helped them feel more relaxed.[9] (The researchers didn't, however, measure cortisol levels.) What's interesting is that this deep breathing technique reduced stress significantly more than doing progressive muscle relaxation beforehand, another popular stress reduction technique.

How do you do this deep breathing technique? Breathe slowly into your belly, also known as diaphragmatic breathing. If you're unfamiliar with diaphragmatic breathing, put a hand on your belly and you should feel it move out as you breathe in. Try to progressively get your breathing down to five breaths per minute, which is what participants were asked to do in the study. Five breaths per minute means that a single breath, from one inhale to the next, takes 12 seconds. You could try a 4-second deep inhale, hold your breath for 4 seconds, then a 4-second exhale.

Normal breathing is typically twelve to eighteen breaths per minute, which might make five breaths per minute seem impossible at first. Try slowing down your breathing and see how low you can go. It helps to do this seated, rather than standing. If you feel lightheaded, definitely stop!

Researchers find that this kind of diaphragmatic breathing is most effective at reducing stress if you do it with a guided audio recording, rather than relying on your own counting.[10] (Perhaps following a recording helps because you're doing it for several minutes, and if left to your own devices, without a guide, your mind could easily wander.) An app definitely comes in handy. Although many meditation apps offer breathwork embedded within meditations, Breathwrk is an app that specializes in breathing. I started using it while I wrote this chapter and there's a free option that has several slow breathing patterns within the "Calming" exercises. I love this app and now use it almost daily. You can first try a breathing exercise for a minute to see if it works for you, then extend it to 5 minutes to enjoy the biggest stress relief benefits.

5. Strike a Pose

When I started researching ways to reduce acute stress, I expected that mindfulness meditation, which we looked at in chapter 6, would be at the top of the list. Indeed, a recent meta-analysis found that, on average, "mindfulness and meditation interventions" significantly reduced cortisol levels and were more effective at lowering levels than talking with a therapist.[11] That's good to know, especially for anyone who can't find or afford therapy. The problem, however, is that most studies saw cortisol levels decrease after eight weeks or more of meditation practice.[12] That doesn't help you with a stressful event that's, say, tomorrow.

But in my search through the mindfulness literature, I did find one practice that can work immediately in a single shot: hatha yoga.

Many studies have found that yoga reduces stress levels, and several have shown that it significantly reduces cortisol.[13] Although you'll reap more stress-reducing benefits if you practice yoga regularly, at the time that I write this, two studies have found that 30 minutes of hatha yoga is enough to lower your cortisol in a stressful situation.[14]

If you're new to yoga, know that there are many types. Hatha yoga, which has been found to be most effective in a single dose, is a slower, more meditative form in which you hold poses for longer periods of time (30 seconds to several minutes), and it often involves mindfulness techniques. (See chapter 5 for more on hatha yoga.) Look online for a beginner's hatha yoga video, which is what the researchers used. You want nonstrenuous poses, such as Tree pose, Child's pose, and Downward Facing Dog.

If you're an experienced yogi, you're either celebrating— (I knew hatha reigned supreme!) or protesting (What about vinyasa yoga? Or yin yoga, which feels very relaxing?). I have one friend who credits Kundalini yoga for her ability to stay in a stressful job for 20+ years. These other types of yoga may also lower stress, especially when done long-term, but in terms of controlled lab studies, it's a single session of hatha yoga that lowered cortisol.

A key limitation in this research is, once again, timing. In these studies, people stood up from their yoga mats and went directly into a stressful situation a few minutes later. Chances are you can't sequester yourself in a guest bedroom at your in-laws' home, do yoga for 30 minutes, then waltz serenely in to dinner, just as everyone's sitting down. Hopefully future researchers will look at whether a single yoga session provides protective effects a few hours later, but for now, it's another potential tool in your kit.

WHAT NOT TO DO

The temptation might be to view the conversations with your mother-in-law as a threat. She does, after all, imply that you work too many hours and that your "neglectful parenting"—yes, she's even used those words—might be why your son is struggling at school. As your visit approaches, you might find yourself anticipating these threatening conversations in exquisite detail and mentally rehearsing your clever rebuttals.

There is a problem, however, with this approach. The problem isn't that you anticipate these crazy-making conversations or even that you rehearse clever comebacks. Practicing what you're going to say can increase your perceived sense of control, and as we saw in the last chapter, increasing perceived control can reduce chronic stress. The problem is that you see the upcoming conversation with her as a threat, and that could easily cancel out any stress-relieving benefits of increased control. Researchers find that if you view a situation as a threat, you're essentially telling yourself that you don't have enough coping resources, and that kind of mental framing, as we're about to see, creates added problems for you.[15] So mentally practing your response is grand, but we need to change your perception of threat. We'll do that in the next section.

---

6. Get Your Butterflies Flying in Formation

So far, we've looked at several strategies to help prevent a full-blown stress response. At their core, these strategies reduce your body's reaction to a difficult situation, which has intuitive and obvious appeal. Try one of these approaches and your jaw probably won't clench as much and your palms won't get as sweaty. More calm, less freak-out. It probably feels like that's as good as it gets.

Now let's consider a completely different approach, one that's less obvious. Instead of calming your body and striving to reduce your stress, our last strategy takes your body's stress response and harnesses it. This one doesn't tamp down the stress your body is experiencing; rather, it converts it into something productive you can use.

This strategy is called "stress reappraisal." In stress reappraisal, you reframe your body's reaction as a sign that your body is mobilizing to help you deal with the stress, rather than as a sign that you can't handle the stress. When you feel your heart beating faster or you break into a sweat, you think, *Interesting—now I'm more laser focused and prepared to respond to this stressful situation*, or *Good, this really matters to me and I'm rising to the occasion*. You're noticing that you feel stressed and you're telling yourself that your body's reaction is a good sign, not a bad one.[16] You're reinterpreting your body's arousal as an adaptive reaction that you can tap into.

An analogy is helpful here. Imagine that you're a skier and you find yourself at the top of a steep, icy mountain.[17] Your only way out is to ski down. Looking down that slippery slope, you're bound to feel high arousal—your breath quickens, your heart pounds, and your stomach might clench up a bit. Avid skiers will interpret this as excitement, believing the trail to be a challenging one that puts their skills to the test but that they can still handle. Novice skiers, however, may interpret their body's reaction as fear, possibly even terror, believing they don't have what it takes to handle a trail this hard. In both cases, they're having the same underlying physiological reaction. The question is how they interpret it.

When you face a stressful situation, you want to view that stressor more like the avid skier who is thinking *It's gonna be hard, but I got this* than the novice skier who is thinking *I can't possibly do this*.

Let's make an important distinction. Stress reappraisal doesn't mean you're telling yourself that it's all okay, that "I like it when my mother-in-law trashes my career. Really, it's refreshing to hear." In that way, it's not quite like the avid skier. You're not pretending you love this. This is hard and stressful, and you genuinely feel your heart pounding. You're not denying any of that.

What you are doing, however, is reinterpreting the stressful situation and your body's reaction to it. Instead of thinking *This is a threat to me* you're thinking *My body's strong reaction is going to help me get through this.* Instead of believing that stress is going to make it hard for you to react, you're seeing the stress as something that is going to *help* you react.

Researchers find that when you reframe stress as something beneficial rather than debilitating, you enjoy a whole host of benefits. First and foremost, you'll have fewer negative emotions during the stressful experience.[18] You won't feel as anxious as a result of your mother-in-law's comments as she makes them, which means you'll probably ruminate on them less later as well. Second, there's a good chance your head will be clearer and you'll react better. College students who had been taught the stress reappraisal technique performed better on subsequent exams than students who hadn't been taught this technique, and adults who used stress reappraisal showed more cognitive flexibility than those who didn't.[19] As we saw in chapter 5 on how to think on your feet, cognitive flexibility is the ability to adjust to changing circumstances, a skill you definitely want if you're responding to jabs about your career or parenting choices. And lastly, with stress reappraisal, you're more likely to notice the upside of the trying experience. Perhaps your mother-in-law only made one comment this time, then she let it go, whereas she usually makes three or four critical comments before she moves on.[20] You wish she hadn't said anything critical, of course, but with stress reappraisal, you're more likely to notice the progress, which can give you hope.

Lisa Feldman Barrett, a neuroscientist at Northeastern University, captures this mental reframing beautifully. Instead of trying to get rid of your nervous butterflies, she says, you're aiming to "make your butterflies fly in formation."[21]

Making That Mental Shift

You might be thinking this is impossible (*Have you* met *my in-laws, Therese?*) but scientists have found ways to help people make this mental shift. The key here, researchers find, is to reframe the stressful situation as a *challenge* you can handle rather than a *threat* you can't.

It's easy to see how an avid skier would achieve that mental state—they've probably practiced on progressively harder slopes so they feel ready to welcome the challenge. But how do you achieve that mindset—I'm ready for this challenge—during stressful situations in which you'd rather not rack up extensive practice?

In research studies, individuals can reach that mental state simply by reading about how their body's natural stress response—such as an increased heart rate and rapid breathing—can actually lead to improved, not worse, performance. As you'll see in the next section, your body's heightened arousal can help you respond firmly, thoughtfully, and diplomatically. Stress, especially acute stress, isn't inherently bad. As we said earlier, cortisol helps your body and mind be more alert, and alert is what you want if you're going to respond carefully to someone's criticisms at a family dinner. You don't want to be as relaxed as you were 5 minutes ago when you were talking about green bean casserole.

The other mental shift you're making is to reassure yourself that yes, your skills do suffice. You can tell yourself, "I've talked about this with my partner / therapist / friends, and I do have the ability to cope with this stressful situation should it arise." As we mentioned earlier, you can identify a few things you can say ahead of time, either to yourself or out loud, to increase your confidence. In my own difficult family moments, I often tell myself, "This is a challenge, not a threat," as a way to reframe what's happening, or "You got this," as a mini pep talk. I also rehearse phrases I can use should a tense moment arise, such as "Let's not talk about that, thank you." Simple, direct, and shuts it down. Plus it helps me feel more prepared for the challenge ahead.

Lisa Feldman Barrett makes the astute observation that when you suddenly feel stressed, it's a sure sign that you've stumbled upon something you value, and that insight can help give you clarity.[22] So you might try being curious, thinking, *Interesting, I must really value this. I wonder what part of this I value?* When I ask myself that, it helps me focus and think more clearly in a stressful moment. I sometimes ask myself this before bed when I'm ruminating on something that happened earlier in the day, and it reliably calms my nerves, allowing me to fall asleep.

I also use reframing when I'm nervous about an upcoming talk. I sometimes can't sleep the night before a big talk, and when I first read this research, I learned to tell myself, "I'm excited, not anxious." Excitement and anxiety often manifest very similarly in the body—a racing mind, an increased heart rate—and even though, deep down, I didn't believe I was excited, I told myself that anyway. I often repeat this mantra, lying in bed, staring at the wall or the ceiling, and most of the time, my body slowly stops freaking out, and I can get back to sleep. You have the ability to help your body reinterpret how it's responding. You're not freaking out; you're working it out.

It's Not Just Silly Semantics—Your Brain and Body Rise to the Occasion

This may all seem like silly semantics, swapping one word for another, but if you see the situation as a challenge, you're telling yourself you have the resources to cope. When you do that, your body and your brain will both react differently.

Let's first look at the body. You know that your heart rate goes up when you're stressed. But is that increased heart rate helping you or not? Researchers find that when you reappraise a stressful situation as a challenge you can handle, your blood vessels open up and your cardiac output goes up.[23] In a nutshell, more blood can reach your brain and your muscles, boosting your ability to think and act. When you interpret a stressful situation as a threat that's too much for you, however, your heart starts pounding, but now your blood vessels constrict, reducing the blood that's sent to your brain and muscles. If you've ever found it hard to think when you feel incredibly stressed and threatened, that's not your "lizard brain" taking over, as so many podcasters like to say. Instead, one thing that's happened is your perception of the threat probably triggered vasoconstriction, so less blood is reaching your brain, making it harder to decide what to do next.

As for how neural activity in your brain changes when you reappraise a stressful situation, that's currently a topic of debate among neuroscientists. One theory that's supported by evidence is that when you try to change the *meaning* of what's happening, certain areas in both the prefrontal cortex and the temporal lobes become more activated.[24]

We've already seen some of the roles that areas in the prefrontal cortex play (as in chapter 9, where we learned the left dorsolateral prefrontal cortex is important for having an action plan), so now let's take a closer look at what the temporal lobes might be doing, which we haven't discussed yet. The temporal lobes, which lie (conveniently enough) just inside your temples,

---

TRY THIS
Your "Manage Intense Moments of Acute Stress" Toolkit

▶ **Hug someone else.** Hug someone who feels safe to you for a full 20 seconds. If you can do this before a stressful event, it should significantly lower your cortisol levels.

▶ **Hug yourself.** Give yourself a soothing self-touch for 20 seconds before a stressful event. Put your hands somewhere that feels comforting to you, such as wrapping your hands around your arms and shoulders or putting one hand on your heart and one on your abdomen.

▶ **Stock up on affection.** The day before or morning of a stressful event, stock up on hugs, kisses, and verbal expressions of affection.

▶ **Slow and deepen your breathing.** Breathe slowly and deeply into your diaphragm for 5 minutes, take a break

---

have a number of important functions, but one of their roles is to understand language and process the meanings of words and images. When you hear the word "yogurt," for example, or see a cup of yogurt in your refrigerator, your temporal lobes light up as you think about what that means. Perhaps you think, "delicious" and "healthy breakfast" (or in the case of my mum, "disgusting" and "someone else better eat that"). But you could reappraise yogurt not as a food but as a beauty product. Applied topically to your face, yogurt can be used to reduce

for a few minutes, then do another 5 minutes of slow, deep breathing. An app can be especially effective here to guide you. Strive to gradually slow your breath to 5 breaths per minute (or 12 seconds from the start of one inhale to the next).

▶ **Do 30 minutes of hatha yoga.** Find a 30-minue introductory hatha yoga video online, and do some yoga before a stressful event.

▶ **Reframe the stressful situation.** Instead of seeing the situation as a threat that you can't cope with, see it as a challenge you can handle. Your body's reaction to stress can actually help you respond more effectively to the situation as long as you see that reaction as beneficial.

acne inflammation or as an exfoliating mask.[25] When you're reappraising yogurt as something you put on your skin rather than something you put in your mouth, your temporal lobes will become activated as you associate new meanings.

Yogurt doesn't create fear or anxiety for most of us, so when you start thinking of new meanings for yogurt, we'd expect much of the increased brain activity to be localized to the temporal lobes, where the meanings of concepts are processed. That activity wouldn't extend to the amygdala, which becomes activated when a concept is of high emotional significance. But the amygdala does light up when you're thinking about something that's emotionally charged, such as when your mother-in-law uses the word "neglectful" to describe your parenting or "selfish" to describe your approach to your career.[26]

Researchers find that when you're reappraising a stressful situation, seeing it as a manageable challenge rather than a humongous threat, your temporal lobes become increasingly activated at the same time that your amygdala becomes less active.[27] So one possibility is that as your temporal lobes are busy assigning a new meaning to what's happening, namely, changing the meaning from *this is a serious threat* to *here's that challenge I thought might happen and I can definitely handle,* your temporal lobes and prefrontal cortex also inhibit your amygdala. After all, you can handle a challenge, so there's no need for your amygdala to overreact. But we do need those temporal lobes to finish assigning a new meaning to what's happening.

So it's not just a silly word game. It's a serious word game, one that helps your body and brain lean into the problem, rather than scramble away from it.

CHAPTER 14

Support Your Better Half

Your partner has been on the job market for months, riding the job search roller coaster. She finds a job that seems like a perfect fit, and her hopes soar, but two weeks later she hasn't heard a thing and her hopes plummet. And then the job posting is removed.

If she only went through this once or twice, it wouldn't be so bad, but she's applied for more than 15 jobs, and the best she's received is an automated email saying they have so many applicants they may never reply. Her mood, even on the weekends, is depressed. She says things like, "What am I doing wrong? I have the skills they're looking for. Why doesn't anyone want me?"

You want to be supportive, so you say it has nothing to do with her. It's probably the economy and the looming recession, you observe sympathetically, and maybe job postings are being removed because companies don't have the funding after all. She looks disheartened and just turns away from you. Then you think of taking some responsibilities off her plate. "How about I make dinner tonight?"

You expect some relief, maybe a mumbled "I would love some soup," or a quick side hug, but instead you get, "Can we just not talk about it?" and a slammed door. Your "helping" clearly isn't.

Social Support: The Boost of Besties

You've stumbled across one of the great paradoxes in social psychology: support sometimes backfires. The intention can be pure and good, and you could go out of your way to do something that, under normal circumstances, she would have welcomed. But now your partner seems more stressed by your gesture, not less.

You're offering what's known as social support. Social support involves providing assistance or comfort to someone else, often to help them cope with a difficult situation.[1] It's the backbone of many close relationships. Social support comes in many forms, from the highly practical—you might shovel your mum's driveway after a heavy snowfall—to the more emotional—you might listen carefully to a friend who feels overwhelmed.

And as your mum or your friend might tell you, social support is often a good thing, if not a great thing. Numerous researchers find that social support is essential for well-being, not just for mental and emotional health but for physical health as well. People who are isolated and receiving little social support are more susceptible to heart disease and breast cancer than people who have close friends and family, and as a result, these lonely people often die younger.[2] Relationships even boost your immune system. In one fascinating study, researchers swabbed people's noses with an active cold virus, and the people with fewer close relationships were more likely to catch the cold, and then became much sicker, than people with many close relationships.[3] (The researchers actually *weighed* their snotty tissues in their trash cans. Fewer friends meant more mucus.)

What's the key takeaway? Healthy relationships keep you healthy.

There's also an emotional strain if you feel no one cares. One study conducted during the first year of Covid found that among people who were self-isolating, individuals who felt they had lots of social support felt less depressed than people who believed they had little support.[4] Everyone in the study lived alone and wasn't going out, but that isolation was much harder on people who felt no one was looking out for them. (If your mum texted you daily for weeks back in 2020, it might have been a tad annoying, but it probably saved your mental health.)

It's not shocking. Most of us function better when we feel supported.

So why isn't the support working with your partner? What's wrong with her?

When Social Support Backfires

Nothing is wrong with her, or with you. The problem, social psychologists would tell you, is that in times of intense,

prolonged stress, certain acts of social support in a close relationship can make the other person feel worse, not better.

Social psychologists have studied how couples respond
to high-stress life events in which the stressful situation lasts
many weeks or months, such as pregnancy and the birth of
a child, studying for the bar exam, or an unsuccessful job
search. Sometimes both partners are struggling equally—
picture the first-time parents who both look sleep-deprived
and are desperate to find some way to soothe their screaming
baby. But often, one partner in the couple is struggling more
than the other—picture the job seeker or stressed-out law
student. Researchers find that when the partner who is coping
says or does something to support the partner who is struggling, the struggling partner sometimes feels mixed emotions
about the help that was thrust upon them or they feel even
more frustrated, depressed, defensive, or overwhelmed than
before.[5] The coping partner is left baffled, wondering, "What
did I say now?"

WHAT DOESN'T WORK

Let's go back to the example at the start of this chapter.
Can you guess what's wrong with the help that's being
offered? Go back and see if anything jumps out at you.

If it all seems harmless, you're not alone. Chances
are we've all said something like "It's out of your control"
or "It has nothing to do with you" to someone we love.
We're trying to make them feel better by pointing out
their predicament isn't their fault.

The problem, however, is that when you said that the economy is probably to blame or that funding might be disappearing, you inadvertently took away some of your partner's control. Most job-seekers are investing a mountain of time and energy into getting a new job. At a cognitive level, your partner is trying to figure out why she isn't getting any nibbles, and if that were her only quandary, your comments would have been perfect. But at an emotional level, she's desperate to figure how she can change her situation, how to have some agency. Should she cut and paste more words from each job description into her resume? Maybe direct message her top contacts on LinkedIn to ask if they're hiring? She wants, maybe even needs, to feel that she can affect the outcome.

And you, in essence, just told her she can't.

Remember what we learned in chapter 12: Perceived control trumps actual control. Don't rob your significant other of her healthy and valuable perception that she can change her plight.

A second problem is that you offered to take away one of her responsibilities: making dinner. In a less stressful time that might be welcome, but right now? That might make her feel more dependent on you and inadvertently signal that either she's less capable or that the family doesn't need her. It will really sting if her ego is tied up in being a good cook.[6] You don't mean to, but you're offering to take away the one meaningful contribution she feels she can make.

So instead of asking, "How about I make dinner tonight?" a better question would be, "I love your cooking, but I'm wondering if this is a night when you'd welcome some time in the kitchen or is this a night when you'd prefer a break?"

It might feel like a slammed door is a pretty bad outcome, but keep it up and things could get much worse. If you repeatedly offer support that your struggling partner doesn't want or need, it begins to wear on your partner's health. In a 10-year study of couples who were living together, there were higher mortality rates among couples in which one partner offered a lot of help that the other partner didn't want.[7] Who was more likely to be dead at the end of the 10 years? It wasn't the partner offering unwanted help, but the partner *receiving* unwanted help. The wrong kind of help truly hurts.

So when, exactly, does social support backfire in a close relationship? Researchers find that it tends to do so under three conditions:

1. You offer help and it reduces the other person's sense of control.
2. You offer a kind of help that isn't wanted or needed.
3. There is already a lot of tension in the relationship (which, unfortunately, is more likely when one partner has been going through prolonged stress).

What Works

The good news is that researchers have found three relatively surefire ways you can offer support to a loved one going through a stressful period.

1. Go Stealth and Resist Taking Credit

The first strategy is a bit sneaky. So far, the support we've talked about is visible support. You offer to make dinner or you blame the economy for her lack of job success. Your partner can see that you're trying to help. And, as we just discussed, that can make your partner feel worse, not better.

But what if you tried invisible support instead? Invisible support is support you offer that your partner might not even consciously notice but that makes her life a little easier and helps her to feel a little better.[8] Perhaps you quietly take out the garbage and recycling, a task she normally does. Perhaps you pick up takeout from her favorite restaurant without announcing, "I got your favorite, California rolls!" Instead, when she sees the food, you simply say, "I was hungry for sushi." Or you might quietly restock her favorite coffee and creamer before she runs out.

Invisible support means offering help that flies under the radar.

Although stressed-out individuals can feel worse when their partner offers obvious support—they might be thinking, *Am I so incompetent that you have to swoop in?*— researchers find that stressed-out individuals typically feel better when their partner offers non-obvious support.[9] It's pretty clever. If you give support that your partner doesn't consciously notice, your partner benefits at multiple levels. Their path through the day is a little easier, they unconsciously feel cared for without the pressure of reciprocating, and they don't feel any of the icky feelings that come with being a person who requires help.

But as you're getting ready for bed that night, you'll need to resist the urge to announce, "You might not have noticed, but I did a good deed today." If the person who is stressed out realizes that they've been helped, the jig is up. Now your invisible support has become visible, and that makes them feel like they can't do crap and that they must be even needier than they realized.

Visible Support Should Still Be Your Normal Go-To

This doesn't mean that all the support you ever offer your spouse should be in stealth mode. Invisible support is most helpful during extended periods of high stress, when your partner is feeling less competent, less capable, and like they'll never achieve their goals.

But in nonstressful times? Visible support saves the day. When it's just life's normal wear and tear, when your partner is complaining about their commute or describing another frustrating email from their sister, then be your normal helpful self. Offer advice on how to respond to that email or coo over the insane traffic. Text her from a coffee shop saying her favorite pumpkin spice lattes are back in season—should you get her one? Researchers find that visible support like this is the backbone of most happy relationships.[10]

2. Be Responsive (Because What Works for You Might Not Work for Someone Else)

You might be thinking, *Therese, what about that 10-year study that found higher death rates in couples where one person offered a lot of support? Doesn't that mean that I should offer less support? I'm not sure I could offer invisible support, so when my partner is struggling, should I just leave them alone?*

No. That study and several that followed found that what's key is responsiveness.

And that brings us to our second strategy. You can offer visible support as long as it's tailored to your partner's preferences. When someone was struggling but had a partner who was responsive to their needs, the struggling person lived longer.[11] It was the stressed-out individuals who essentially said, "My partner doesn't really try to understand me but spends a lot of time trying to help me," who had poor health and shorter lives.

Concrete advice about how to be responsive is a little hard to give because the whole point is that your help needs to be responsive to what your partner needs. Needs vary. One person might be eternally grateful if you said, "I'll take the kids to my parents for the weekend so you can be alone to work," whereas another might become even more disheartened at the thought of bumping around the house by themselves for 48 hours, checking their phone and inbox every 10 minutes.

You might have a hard time gauging what your partner wants or needs right now, so pick a good time to ask them. But don't do it when they're struggling. When your partner is in a neutral or good mood, find out what would feel supportive. Say something like, "I can see this is a stressful time and I care so much about you. I know you've got the strength and resilience to get through this, but I want to help if I can. So what would real support look like? What could I start doing or stop doing that would make your life a little easier?"

3. A Warm Touch Works Wonders

Is there a nueroscience-backed way to support a stressed-out loved one?

There is, and the best part is that it's simpler than either of the other approaches. Touch.

First, touch is nature's stress reliever.[12] As we saw in chapter 13, affectionate touch from someone you care about reduces the anxiety and pressure you're feeling. In studies in which participants kept daily diaries, on days when affection went up, their stress levels went down.[13] The calming effects of touch lasted well into the next day—when individuals enjoyed touch on one day, they found the next day's events less upsetting and easier to deal with.

But perhaps even more importantly, touch from someone you care about reduces activity in brain regions associated with negative emotions. An international team of neuroscientists was curious how different forms of social support would affect the brain activity of stressed individuals.[14] They had women bring a close female friend into the lab, and while one woman was getting ready to go into the scanner, the close friend was instructed how to support her.

The anxiety-provoking situation was one of rejection, in which the person in the scanner had been made to feel as though they weren't good enough or liked for some reason (much like how your job-searching partner is feeling rejected). Some of the friends outside the scanner were instructed to explain away the rejection so that it didn't feel so personal. They were given things to say to reassure their friend and blame the rejection on things outside their control. Other friends were instructed to gently touch the hand of their friend but say nothing.

The data are interesting and a bit surprising. First, the people who had been touched reported feeling better than the people who had been reassured. That's not surprising, given what you've learned in this chapter and the previous chapter on stress, but what *is* surprising is that the reassurances backfired. Participants who had a friend reassure them actually felt *worse* than people who came to the lab without a friend. It's not clear why the reassurance backfired, but perhaps it's because the reassurance suggested there was nothing anyone could do. That felt worse than no comment at all.

Second, the researchers examined what happened in the brain in response to different kinds of support. People who received a brief comforting touch on the hand showed reduced activity in the anterior insula, a region that, as we learned in chapter 8, becomes more active when a person is feeling emotional pain. A brief touch accomplished what the reassurance was supposed to do: It made the rejection hurt less. The people who were verbally reassured by a friend, on the other hand, showed no change in activity in their agitated anterior insula. Touch calmed the nervous system, reassurances did not.

I love this study because my go-to strategy, before reading this research, was to try to comfort a friend by explaining away a rejection. "You know," I'd say, "X is likely the real reason they aren't getting back to you. It probably has nothing to do with you." I swear I've said these exact words to friends on the job market. At the time, "nothing to do with you" felt insightful and helpful. The research makes it clear, though—maybe I was insightful, but that's not the same as helpful. "Nothing to do with you" is heard as "Nothing you can do," and as we saw in chapter 12, perceived control is crucial to coping with tense situations.

So when your partner is stressed, your simple go-to solution can be a reassuring touch. Put your head on their shoulder, rub their back, wrap them in a gentle, silent hug.

One important caveat: You need to be respectful of how your partner responds to touch. Maybe they have some past trauma from which they're healing and a long, unexpected hug is going to trigger alarm bells. Use good judgment about what kind of touch your partner would welcome.

TRY THIS
Your "Support Your Better Half" Toolkit

▶ **Provide invisible support.** If your partner has been stressed for a long time and is beginning to feel like they have little control over their situation, nonobvious, invisible support will probably be more appreciated than obvious, visible support. Do things that make your partner's day easier, like taking out the trash, without taking credit.

▶ **Be responsive.** Visible support can still be appreciated as long as it's responsive and tailored to your partner's needs. When your partner is in a neutral or good mood, ask, "What would support look like right now? What would you like me to start or stop doing?"

▶ **Touch works wonders.** Gentle, affectionate touch reduces stress and can be more beneficial than offering verbal reassurances. Try hugging your partner or putting your hand on their arm when they're upset.

A Parting Word to the Wise

Congratulations! You're much sharper now than when you first picked up this book. You have more practical know-how about your brain, and you've learned how to fine-tune certain parts of your nervous system so you can become more of the person you'd like to be.

But being sharp isn't just about reading. It's about *doing*. Even if you don't currently face any of the challenges described in this book, pick two strategies and put them into practice. Start today. The Appendix that follows tells you where you can find the quickest strategies in this book. If it's a weekday and you have no time, start with a two-minute strategy. If it's the weekend and you have a little more time, pick a strategy from the "5 minutes or less" list.

Why start now? The reason is simple: Whatever strategies you start practicing now will be tucked away in your handy toolkit. You'll have learned them before you're crunching for a deadline or walking into an important meeting. When you do desperately need that tool, you'll reach for it with ease, rather than fumbling around, getting frustrated.

I can use myself as an example. It was Christmas 2023 and I had recently completed a full draft of this book. My husband and I pulled up to visit his parents around 2:00 p.m. on Christmas afternoon, and I felt myself getting incredibly anxious, as I have for years now. I didn't want to go in, I wanted to pick a fight with my husband, I wanted to pretend I was sick so I could excuse myself for the day.

But I was able to calm myself down more quickly and easily than I ever have before, thanks to what I learned while working on this book. I used the techniques on handling acute stress and the breathing strategies I've described, both in the car and throughout the afternoon.

Did I remember ahead of time that I would need those tools that day? I did not. But I was already practicing those techniques regularly, so when the time came, I knew exactly what to do. We walked in, I stayed chill, and we had a lovely Christmas. I now know that stressful situations are a little more predictable (at least my reaction to them will be), and that fact alone is an incredible balm in an unpredictable world.

So start practicing the strategies now, before you need them, so that, like me, you'll have them at your fingertips.

I Want to Help You 2.0

If you're reading this, you've made it to the final pages of this book, and you're committed to being as sharp and as keen as you can be. If I could hug you, I would. Because we've made it this far together, I want to offer a little more.

I wish this book could be all that you'll ever need to be the best version of you. Book read, brain improved. Done and done. Perfection achieved.

But neuroscience is constantly changing. According to one recent search, approximately 2,000 to 3,000 new articles are published in peer-reviewed neuroscience journals every single month. That's a lot of new information. Most of that research is building upon what we already know, like adding another layer to an existing pyramid. New strategies are discovered, old ones are improved upon.

Occasionally, however, there are game-changing research findings. Sometimes there is a finding that turns everything on its head, the one that becomes the main buzz at a conference. It's important to know this because although I've tried to present the most up-to-date research I can, there are bound to be new discoveries that challenge something I've said.

So when you have a minute, please go to my website, theresehuston.com, where I'll explore some of these important new discoveries and strategies. My intent is to stay on top of the ever-evolving research in order to provide the best new practices and most interesting findings you can add to your growing toolkit.

You and Your Potential

Before you close this book, decide which practices you are going to try today. Seriously, what will you do before you go to bed tonight? Maybe you'll grab your headphones and listen to some binaural beats to improve your focus, or you'll do a 20-second self-hug to reduce your stress. Perhaps you'll play some music that gives you chills to rev up your motivation, or you'll register for a workshop on self-compassion.

I can honestly say that since I started employing many of these strategies, I live a better life. I waste less time when I sit down to my computer each morning, and I wrap up my workday earlier, often by 4:15 or 4:30, because I've been so productive. I've replaced my weird doctor with someone I like and trust, I've added about 40 minutes of sleep each night, and my heart rate variability has increased by 30%. I'm not special or heroic, and I don't have a nanny or a personal chef or a trust fund to ease life's challenges. I'm just improving my life and getting sharper, one strategy at at time. I didn't live a bad life three years ago when I started researching this book, but man, I live a really good life today.

Now it's your turn.

Be bold and pick two of the strategies we've discussed and try them before the day is over. Then build from there, and watch your life transform. Because whether you realize it or not, you just learned how to reach your fullest potential by changing how you spend 5–30 minutes each day. You have to put in the work, but I think you'll find it fairly easy, and, I hope, fun! So go surprise yourself. Your sharpest self is waiting.

Acknowledgments

When I was a kid and dreamed of being a writer, I would save small slips of paper in my big wooden jewelry box with the names of future characters written on them. My grandpa had made me the jewelry box, so it held a place of honor in my bedroom, but I preferred collecting names over necklaces. Whenever I came across a name I liked, I'd scribble it down and tuck it away to use in a book someday. Nick Carrocio. Jasmin Sinclair.

I thought I'd write fiction.

I thought wrong.

But what I especially got wrong was believing that the melodic names of fictional people would elevate my writing. Silly me. It's the real people with real names who make my books worth reading.

Here are the names of the wonderful people who helped me write this book. More importantly, here's a glimpse of what they contributed:

Vision — To my literary agent, Lindsay Edgecombe, who saw the potential in my early ramblings and championed me and my work tirelessly.

Rigor — To my early mentors, Marlene Behrmann, Cameron Carter, Jonathan Cohen, and James McClelland, who taught me how to think like a neuroscientist.

Astuteness — To my first editor on this book, Daniela Rapp, who sharpened my thinking and kept me grounded in news one could use.

Enthusiasm — To my second editor, Annie Chagnot, who picked up this manuscript and ran with it, making my words flow and me feel triumphant.

Persistence — To the editorial team at Mayo Clinic Press, especially Jenny Krueger, Nina Wiener, and Dana Noble, who have stepped up to fill many needs.

Thoroughness — To my production editor, Alan Bradshaw, for demonstrating that "clear and comprehensive" can be superpowers.

Creativity — To the book's designer, Amanda Knapp, and the book's typesetter, Andy Berry, who made this book a joy to look at.

Attentiveness — To the book's copy editor, James Bradshaw, and proofreader, Rima Weinberg, who caught all the little mistakes that annoy careful readers.

Diligence — To the medical reviewer, John Henley, who generously verified that the book's science is sound.

Encouragement — To my friends and family who asked supportive questions about this book over tea, lunch, walks, and long phone calls, including Jamie Adaway, Mark Cohan, Irina Drozdova, David Green, Maria Farmer, Katie Foster, Janelle Hillhouse, Meghan and Chad Lyle, Jacquelyn Miller, Katherine Raichle, Linda Selig, and Giannina Silverman.

Insight — A special call-out to my friend, Elizabeth Haydn-Jones, who read early chapters and gave me playful ideas and insights that found their way into my thinking.

Support — To my mom, Karen Gee, for cheering me on to pursue neuroscience, from grad school (when she bought me a technical book I couldn't afford) all the way to now.

Love — To my husband, Jonathan, who left a job to bring me back to my favorite place to write, who brainstorms whenever I ask, and who gives me a warm place to land on days I overdo it (which, he'll say with an affectionate grin, is often).

And if you've helped me with this book and I have somehow forgotten, I still thank you. So many people nudged me in the right direction, and I wish I'd saved your names on little slips of paper. Next jewelry box.

Quickest Strategies for a Better You

If you're pressed for time (and let's be honest, who isn't?), this Appendix is for you. I've identified almost 30 strategies that take 5 minutes or less, and I've organized them so that you can jump right to the chapter you need.

Be patient. Doing the strategy itself might only take 2 minutes, but it will probably take longer to read about it, understand it well enough to do it correctly, and gather any materials you might need. And if you find that a particular strategy takes 5 or 6 minutes, not 2, that doesn't mean you're doing it incorrectly. I've given my best estimates, but we're all different.

STRATEGY	CHAPTER, TOPIC	PAGE
Time Required: Two Minutes or Less		
Drink Tea	Chapter 1, Get Focused	32
Grab Your Headphones or Airpods (This will only take 2 minutes to launch, but you should listen longer while you're working)	Chapter 1, Get Focused	34
Stare, But Not into Space	Chapter 1, Get Focused	41
Eat Chicken or Edamame	Chapter 2, Get Creative	50
Drink a Big Cup of Joe	Chapter 3, Get Motivated	71
Test Yourself Before You Rest Yourself	Chapter 6, Learn More and Learn it Fast	124
Find Your Motive	Chapter 8, Relate More	151
Slow Your Exhale	Chapter 10, Make Better Decisions	191
Ask Yourself, "What Could I Gain?"	Chapter 12, Take Charge of Chronic Stress	233
Slow Your Breathing	Chapter 12, Take Charge of Chronic Stress	235
Hug Someone Else	Chapter 13, Manage Intense Moments of Acute Stress	255
Hug Yourself	Chapter 13, Manage Intense Moments of Acute Stress	256
A Warm Touch Works Wonders	Chapter 14, Support Your Better Half	265

STRATEGY	CHAPTER, TOPIC	PAGE
Time Required: Five Minutes or Less		
Go for a Walk (⅛ mile)	Chapter 2, Get Creative	46
Remember a TV Show	Chapter 2, Get Creative	54
Find the Bigger Goal	Chapter 3, Get Motivated	63
Listen to Music That Moves You	Chapter 3, Get Motivated	68
Do the Right Kind of Self-Affirmation	Chapter 3, Get Motivated	73
Picture the Process	Chapter 4, Accomplish More	83
Scare Yourself Smart	Chapter 6, Learn More and Learn it Fast	126
Identify Baby Steps	Chapter 7, Make Fewer Mistakes	140
Blur Group Boundaries	Chapter 9, Be More Fair and Less Biased	178
Put Yourself in the Driver's Seat	Chapter 10, Make Better Decisions	189
Try a "Look Back"	Chapter 10, Make Better Decisions	190
Name It to Tame It	Chapter 12, Take Charge of Chronic Stress	232
Stock Up on Affection	Chapter 13, Manage Intense Moments of Acute Stress	243
Reframe the Stressful Situation	Chapter 13, Manage Intense Moments of Acute Stress	256
Be Responsive	Chapter 14, Support Your Better Half	264

Notes

Introduction

1.   Sasha J. Davies et al., "Cognitive Impairment During Pregnancy: A Meta-analysis," *Medical Journal of Australia* 208, no. 1 (2018): 35-40, https://pubmed.ncbi.nlm.nih.gov/29320671/.

2.   Alexander Bystritsky et al., "Brain Circuitry Underlying the ABC Model of Anxiety," *Journal of Psychiatric Research* 138 (2021): 3-14, https://www.sciencedirect.com/science/article/abs/pii/S0022395621001801.

3.   Where did the 10% myth come from? Many attribute it to William James, the father of psychology, who said, in *The Energies of Men,* we make use "of only a small part of our possible mental and physical resources," but I think popular culture is the real culprit. Movies like *Limitless* and *Lucy* keep the myth alive and well. Kelly Macdonald et al., "Dispelling the Myth: Training in Education or Neuroscience Decreases but Does Not Eliminate Beliefs in Neuromyths," *Frontiers in Psychology* 8 (2017): 1314, https://www.frontiersin.org/articles/10.3389/fpsyg.2017.01314/full. William James, "The Energies of Men," first published in *Science, N.S.* 25, no. 635 (1907): 321-332, https://psychclassics.yorku.ca/James/energies.htm.

4.   John Henley, quoted in Robynne Boyd, "Do People Only Use 10 Percent of Their Brains?" *Scientific American* 7 (2008), https://www.scientificamerican.com/article/do-people-only-use-10-percent-of-their-brains/.

Chapter 1

1. Richard Carciofo et al., "Chronotype and Time-of-Day Correlates of Mind Wandering and Related Phenomena," *Biological Rhythm Research* 45, no. 1 (2014): 37-49.

2. Marcel Adam Just, Timothy A. Keller, and Jacquelyn Cynkar, "A Decrease in Brain Activation Associated with Driving when Listening to Someone Speak," *Brain Research* 1205 (2008): 70-80. This is a famous study of how driving while listening carefully to another person speaking impairs driving performance. The increase in error rates was calculated using the data provided on p. 72: road maintenance errors increased from an average of 8.7 when driving undistracted to an average of 12.8 when driving while listening, for an increased error rate of 47%.

3. Kevin P. Madore et al., "Memory Failure Predicted by Attention Lapsing and Media Multitasking," *Nature* 587, no. 7832 (2020): 87-91.

4. David M. Sanbonmatsu et al., "Who Multi-Tasks and Why? Multi-Tasking Ability, Perceived Multi-Tasking Ability, Impulsivity, and Sensation Seeking," *PLoS One* 8, no. 1 (2013): e54402.

5. Antonella Samoggia and Tommaso Rezzaghi "The Consumption of Caffeine-Containing Products to Enhance Sports Performance: An Application of an Extended Model of the Theory of Planned Behavior," *Nutrients* 13, no. 2(2021):344.

6. For specific research on the attention-boosting benefits of caffeine plus l-theanine in green tea, see Christina Dietz and Matthijs Dekker, "Effect of Green Tea Phytochemicals on Mood and Cognition," *Current Pharmaceutical Design* 23, no. 19 (2017): 2876-2905. For an analysis of l-theanine and caffeine more generally, see Chanaka N. Kahathuduwa et al., "Acute Effects of Theanine, Caffeine and Theanine–Caffeine Combination on Attention," *Nutritional Neuroscience* 20, no. 6 (2017): 369-377.

7. Jackson L. Williams et al., "The Effects of Green Tea Amino Acid L-theanine Consumption on the Ability to Manage Stress and Anxiety Levels: A Systematic Review," *Plant Foods for Human Nutrition* 75 (2020): 12-23.

8. Chanaka Kahathuduwa et al., "L-theanine and Caffeine Improve Sustained Attention, Impulsivity and Cognition in Children with Attention Deficit Hyperactivity Disorders by Decreasing Mind Wandering (OR29-04-19)," *Current Developments in Nutrition* 3, Suppl. 1 (2019): nzz031-OR29. See also Chanaka N. Kahathuduwa et al., "Effects of L-theanine–Caffeine Combination on Sustained Attention and Inhibitory Control Among Children with ADHD: A Proof-of-Concept Neuroimaging RCT," *Scientific Reports* 10, no. 1 (2020): 13072.

9. Ethan M. McCormick and Eva H. Telzer, "Contributions of Default Mode Network Stability and Deactivation to Adolescent Task Engagement," *Scientific Reports* 8, no. 1 (2018): 18049.

10. Emma K. Keenan et al., "How Much Theanine in a Cup of Tea? Effects of Tea Type and Method of Preparation," *Food Chemistry* 125, no. 2 (2011): 588-594.

11. How about almond milk? Or oat milk? At the time I write this, I can't find any research on whether plant-based milks disrupt the l-theanine content in tea. Dairy milk binds to the l-theanine, making it harder for your body to absorb. As for how much dairy milk is too much, the researchers found that more than 3 tablespoons added to a 6.5 oz teacup brought the l-theanine levels to near zero.

12. Sarah Krull Abe and Manami Inoue, "Green Tea and Cancer and Cardiometabolic Diseases: A Review of the Current Epidemiological Evidence," *European Journal of Clinical Nutrition* 75, no. 6 (2021): 865-876.

13. James D. Lane et al., "Binaural Auditory Beats Affect Vigilance Performance and Mood," *Physiology & Behavior* 63, no. 2 (1998): 249-252.

14. For a literature review on how binaural beats improve focused attention, see Sandhya Basu and Bidisha Banerjee, "Potential of Binaural Beats Intervention for Improving Memory and Attention: Insights from Meta-analysis and Systematic Review," *Psychological Research* (2022): 1-13. For research on how binaural beats reduce mind wandering, see Ulrich Kirk et al., "On-the-Spot Binaural Beats and Mindfulness Reduces Behavioral Markers of Mind Wandering," *Journal of Cognitive Enhancement* 3 (2019): 186-192.

15. There's some debate as to the maximum frequency of brain waves. Some report that gamma waves top out at 80 Hz, others say 100 is the magic number, and a select few claim they can go as high as 150 for select individuals.

16. Marie T. Banich and Rebecca J. Compton, *Cognitive Neuroscience* (Cambridge University Press, 2018).

17. Hessel Engelbregt et al., "Effects of Binaural and Monaural Beat Stimulation on Attention and EEG," *Experimental Brain Research* 239, no. 9 (2021): 2781-2791.

18. Matthew K. Robison et al., "The Effect of Binaural Beat Stimulation on Sustained Attention," *Psychological Research* 86, no. 3 (2022): 808-822.

19. Magda Jordão et al., "Meta-analysis of Aging Effects in Mind Wandering: Methodological and Sociodemographic Factors," *Psychology and Aging* 34, no. 4 (2019): 531, https://drive.google.com/file/d/10DBLJZBk0CY1p898Ck946bk-lLuo0otM/view.

20. Researchers have different perspectives on when the ability to focus begins to decline. It seems to depend on how they measure and define "focus," "attention," and one's ability to concentrate and ignore distractions. For research on how focus peaks in one's early thirties, see Timothy A. Salthouse, "Selective Review of Cognitive Aging," *Journal of the International Neuropsychological Society* 16, no. 5 (2010): 754-760, https://www.ncbi.nlm.nih.gov/pmc/articles/PMC3637655/. For research on how focus peaks in one's early forties, see Francesca C. Fortenbaugh et al., "Sustained Attention Across the Life Span in a Sample of 10,000: Dissociating Ability and Strategy," *Psychological Science* 26, no. 9 (2015):

1497-1510, https://journals.sagepub.com/doi/10.1177/0956797615594896. For research on how the big drop happens in one's sixties, see Erika Borella, Barbara Carretti, and Rossana De Beni, "Working Memory and Inhibition Across the Adult Life-Span," *Acta Psychologica* 128, no. 1 (2008): 33-44, https://pubmed.ncbi.nlm.nih.gov/17983608/.

21. Lixia Yang, Kesaan Kandasamy, and Lynn Hasher, "Inhibition and Creativity in Aging: Does Distractibility Enhance Creativity?," *Annual Review of Developmental Psychology* 4 (2022): 353-375, https://www.annualreviews.org/doi/pdf/10.1146/annurev-devpsych-121020-030705.

22. Janice Colleen McMurray, "Binaural Beats Enhance Alpha Wave Activity, Memory, and Attention in Healthy-Aging Seniors" (PhD diss., University of Nevada, Las Vegas, 2006).

23. Annette Shamala Arokiaraj, Rozainee Khairudin, and WS Wan Sulaiman, "The Impact of a Computerized Cognitive Training on Healthy Older Adults: A Systematic Review Focused on Processing Speed and Attention," *International Journal of Academic Research in Business and Social Sciences* 10, no. 11 (2020): 645-685, https://www.semanticscholar.org/paper/The-Impact-of-a-Computerized-Cognitive-Training-on-Arokiaraj-Khairudin/b961776a80db ca37967d97d470fe23d1d81448e5?p2df.

24. Yi-Jung Lai and Kang-Ming Chang, "Improvement of Attention in Elementary School Students Through Fixation Focus Training Activity," *International Journal of Environmental Research and Public Health* 17, no. 13 (2020): 4780.

25. Barbara Franca Haverkamp et al., "Effects of Physical Activity Interventions on Cognitive Outcomes and Academic Performance in Adolescents and Young Adults: A Meta-analysis," *Journal of Sports Sciences* 38, no. 23 (2020): 2637-2660.

26. Emily Balcetis, *Clearer, Closer, Better: How Successful People See the World* (New York: Ballantine Books, 2021).

Chapter 2

1. Mark A. Runco and Garrett J. Jaeger, "The Standard Definition of Creativity," *Creativity Research Journal* 24, no. 1 (2012): 92-96.

2. Martin Meinel et al., "Designing Creativity-Enhancing Workspaces: A Critical Look at Empirical Evidence," *Journal of Technology and Innovation Management* 1, no. 1 (2017). For specific work on the effects of plants and nature, see Kathryn J. H. Williams et al., "Conceptualising Creativity Benefits of Nature Experience: Attention Restoration and Mind Wandering as Complementary Processes," *Journal of Environmental Psychology* 59 (2018): 36-45.

3. Christian Rominger et al., "Acute and Chronic Physical Activity Increases Creative Ideation Performance: A Systematic Review and Multilevel Meta-

analysis," *Sports Medicine-Open* 8, no. 1 (2022): 1-17.

4. Christian Rominger et al., "Step-by-Step to More Creativity: The Number of Steps in Everyday Life Is Related to Creative Ideation Performance," *American Psychologist* (2023), advance online publication, https://doi.org/10.1037/amp0001232.

5. Chun-Yu Kuo and Yei-Yu Yeh, "Sensorimotor-Conceptual Integration in Free Walking Enhances Divergent Thinking for Young and Older Adults," *Frontiers in Psychology* 7 (2016): 1580.

6. Supriya Murali and Barbara Händel, "Motor Restrictions Impair Divergent Thinking During Walking and During Sitting," *Psychological Research* 86, no. 7 (2022): 2144-2157.

7. Marily Oppezzo, "Want to Be More Creative? Go for a Walk," TED, April 2017, video, 5:15, https://www.ted.com/talks/marily_oppezzo_want_to_be_more_creative_go_for_a_walk?language=en.

8. Matthijs Baas, Carsten K. W. De Dreu, and Bernard A. Nijstad, "A Meta-analysis of 25 Years of Mood-Creativity Research: Hedonic Tone, Activation, or Regulatory Focus?," *Psychological Bulletin* 134, no. 6 (2008): 779.

9. Jihae Shin and Adam M. Grant, "When Putting Work Off Pays Off: The Curvilinear Relationship Between Procrastination and Creativity," *Academy of Management Journal* 64, no. 3 (2021): 772-798.

10. Ibid.

11. Lixia Yang, Kesaan Kandasamy, and Lynn Hasher, "Inhibition and Creativity in Aging: Does Distractibility Enhance Creativity?," *Annual Review of Developmental Psychology* 4 (2022): 353-375, https://www.annualreviews.org/doi/pdf/10.1146/annurev-devpsych-121020-030705.

12. Hikaru Takeuchi et al., "Regional Gray Matter Volume of Dopaminergic System Associate with Creativity: Evidence from Voxel-Based Morphometry," *Neuroimage* 51, no. 2 (2010): 578-585.

13. Lorenza S. Colzato, Annelies M. de Haan, and Bernhard Hommel, "Food for Creativity: Tyrosine Promotes Deep Thinking," *Psychological Research* 79 (2015): 709-714.

14. Soghra Akbari Chermahini and Bernhard Hommel, "The (B) Link Between Creativity and Dopamine: Spontaneous Eye Blink Rates Predict and Dissociate Divergent and Convergent Thinking," *Cognition* 115, no. 3 (2010): 458-465.

15. Bryant J. Jongkees et al., "Effect of Tyrosine Supplementation on Clinical and Healthy Populations Under Stress or Cognitive Demands—A Review," *Journal of Psychiatric Research* 70 (2015): 50-57.

16. For research on how increased dopamine levels lead to impulse control problems, see Marie Grall-Bronnec et al., "Dopamine Agonists and Impulse Control Disorders: A Complex Association," *Drug Safety* 41 (2018): 19-75.

17. Kevin P. Madore et al., "Neural Mechanisms of Episodic Retrieval Support Divergent Creative Thinking," *Cerebral Cortex* 29, no. 1 (2019): 150-166. For a more recent study exploring how this episodic retrieval activity improves creative writing, see Ruben D. I. van Genugten et al., "Does Episodic Retrieval Contribute to Creative Writing? An Exploratory Study," *Creativity Research Journal* 34, no. 2 (2022): 145-158.

18. Daniel L. Schacter, Daniel T. Gilbert, and Daniel M. Wegner, *Psychology* (New York: Macmillan, 2009).

19. Morris Moscovitch, "Episodic Memory and Beyond: The Hippocampus and Neocortex in Transformation," *Annual Review of Psychology* 67 (2016): 105-134.

20. According to one biographer, this comment was actually made by Mark Twain's wife, Olivia Langdon Clemens, in a discussion on a train about whether it's possible to copyright ideas, but, rather ironically, the quote is often credited to Twain. Albert Bigelow Paine, *Mark Twain: A Biography; the Personal and Literary Life of Samuel Langhorne Clemens*, vol. 3. (New York: Harper & Brothers, 1912), quoting a conversation between Mark Twain and Mrs. Clemens on p. 1343.

21. Madore et al., "Neural Mechanisms of Episodic Retrieval," and Van Genugten et al., "Episodic Retrieval and Creative Writing." Note that in these studies, the researchers had all participants watch the same video and then recall details from that video. I've modified their procedure a bit to fit your real life. The benefits should be the same.

Chapter 3

1. Jürgen Wegge and S. Alexander Haslam, "Improving Work Motivation and Performance in Brainstorming Groups: The Effects of Three Group Goal-Setting Strategies," *European Journal of Work and Organizational Psychology* 14, no. 4 (2005): 400-430, https://www.tandfonline.com/doi/abs/10.1080/13594320500349961?role=button&needAccess=true&journalCode=pewo20.

2. Edwin A. Locke and Gary P. Latham, "Building a Practically Useful Theory of Goal Setting and Task Motivation: A 35-Year Odyssey," *American Psychologist* 57, no. 9 (2002): 705, https://psycnet.apa.org/doiLanding?doi=10.1037%2F0003-066X.57.9.705.

3. Dr. Anna Lembke, "Tools to Manage Dopamine and Improve Motivation & Drive," interview by Andrew Huberman, *The Huberman Lab,* October 6, 2022, https://hubermanlab.com/tools-to-manage-dopamine-and-improve-motivation-and-drive/#:~:text=%E2%80%9CDopamine%20is%20about%20wanting%2C%20not,she%20is%20100%25%20correct.

4. Andrew Westbrook et al., "Dopamine Promotes Cognitive Effort by Biasing the Benefits Versus Costs of Cognitive Work," *Science* 367, no. 6484 (2020): 1362-1366, https://www.ncbi.nlm.nih.gov/pmc/articles/PMC7430502/.

5. Valorie N. Salimpoor et al., "Anatomically Distinct Dopamine Release During Anticipation and Experience of Peak Emotion to Music," *Nature Neuroscience* 14, no. 2 (2011): 257-262, http://audition.ens.fr/P2web/eval2011/BT _Salimpoor2011.pdf.

6. Ibid.

7. Petr Srámek et al., "Human Physiological Responses to Immersion into Water of Different Temperatures," *European Journal of Applied Physiology* 81 (2000): 436-442, https://wildaufleben.at/studien/Human%20physiological%20 responses%20to%20immersion%20into%20water%20of%20different%20 temperatures_Sramek%201999.pdf.

8. The closest study I can find is one that found that after 30 seconds in a cold shower, done daily for 30 consecutive days, adults missed fewer days of work because of illness in the 90 days that followed. But no blood tests were conducted and it was 30 consecutive days, not occasional cold showers. The only other study I can find that looked at brief cold showers was published back in 1964 and found no significant changes in adrenal catecholamines (they didn't even test dopamine). For the study that found fewer sick days, see Geert A. Buijze et al., "The Effect of Cold Showering on Health and Work: A Randomized Controlled Trial," *PLoS One* 11, no. 9 (2016): e0161749.

9. N. D. Volkow et al., "Caffeine Increases Striatal Dopamine D2/D3 Receptor Availability in the Human Brain," *Translational Psychiatry* 5, no. 4 (2015): e549-e549, https://www.nature.com/articles/tp201546.

10. Caffeine levels were obtained from the following database (April 2023): https://www.caffeineinformer.com/the-caffeine-database.

11. "Get Smart About Caffeine," National Consumers League (blog), March 1, 2016, https://nclnet.org/caffeine_facts/#:~:text=According%20to%20the%20 Dietary%20Guidelines,men%20ages%2051%2D70.

12. Christopher N. Cascio et al., "Self-Affirmation Activates Brain Systems Associated with Self-Related Processing and Reward and Is Reinforced by Future Orientation," *Social Cognitive and Affective Neuroscience* 11, no. 4 (2016): 621-629, https://www.ncbi.nlm.nih.gov/pmc/articles/PMC4814782/.

13. Geoffrey L. Cohen and David K. Sherman, "The Psychology of Change: Self-Affirmation and Social Psychological Intervention," *Annual Review of Psychology* 65 (2014): 333-371.

Chapter 4

1. Thomas L. Webb and Paschal Sheeran, "Does Changing Behavioral Intentions Engender Behavior Change? A Meta-analysis of the Experimental Evidence," *Psychological Bulletin* 132, no. 2 (2006): 249, https://psycnet.apa.org/record /2006-03023-004.

2. Heather Barry Kappes and Gabriele Oettingen, "Positive Fantasies About Idealized Futures Sap Energy," *Journal of Experimental Social Psychology* 47, no. 4 (2011): 719-729, https://www.sciencedirect.com/science/article/abs/pii /S002210311100031X?casa_token=Rn_C9c53FzYAAAAA:UHyuGmmel6y HRfjMcixRz85pHf89KKMXKOpe9cmlWjIhvoU9XFvOtrCkElwe38VrU4i Ja0OGDg.

3. Lien B. Pham and Shelley E. Taylor, "From Thought to Action: Effects of Process- Versus Outcome-Based Mental Simulations on Performance," *Personality and Social Psychology Bulletin* 25, no. 2 (1999): 250-260, https://journals .sagepub.com/doi/pdf/10.1177/0146167299025002010?casa_token=rIFs _iaQgdAAAAAA:4kQT7vXnbSdslw-BwUA0hDAtdkFH6a9gwXZG9aeKG _f9QLO36QZUZ6AYvW8sfEeqMdYIYNmV_asa.

4. Gabriele Oettingen, Doris Mayer, and Sam Portnow, "Pleasure Now, Pain Later: Positive Fantasies About the Future Predict Symptoms of Depression," *Psychological Science* 27, no. 3 (2016): 345-353, https://s18798.pcdn.co/motivationlab /wp-content/uploads/sites/6235/2019/02/oettingen-et-al-2016-pleasure-now-pain -later.pdf.

5. Kathy D. Gerlach et al., "Future Planning: Default Network Activity Couples with Frontoparietal Control Network and Reward-Processing Regions During Process and Outcome Simulations," *Social Cognitive and Affective Neuroscience* 9, no. 12 (2014): 1942-1951, https://academic.oup.com/scan/article/9/12/1942/1615827.

6. Pham and Taylor, "From Thought to Action."

7. Gerlach et al., "Future Planning."

8. Thomas Baumgartner et al., "Dorsolateral and Ventromedial Prefrontal Cortex Orchestrate Normative Choice," *Nature Neuroscience* 14, no. 11 (2011): 1468-1474, https://www.zora.uzh.ch/id/eprint/50019/1/plugin-Baumgartner_Fehr _HumanBrainMapping2011_Manuscript.pdf.

9. Pureheart Ogheneogaga Rikefe, "Effect of Objectives and Key Results (OKR) on Organisational Performance in the Hospitality Industry," *International Journal of Research Publications* 91, no. 1 (2021): 185-195, https://www .researchgate.net/profile/Pureheart-Irikefe/publication/357521938_Effect_of _Objectives_and_Key_Results_OKR_on_Organisational_Performance_in_the _Hospitality_Industry/links/61ef2a5b8d338833e392bbfc/Effect-of-Objectives -and-Key-Results-OKR-on-Organisational-Performance-in-the-Hospitality -Industry.pdf.

10. Tracy Epton, Sinead Currie, and Christopher J. Armitage, "Unique Effects of Setting Goals on Behavior Change: Systematic Review and Meta-analysis," *Journal of Consulting and Clinical Psychology* 85, no. 12 (2017): 1182, https://psycnet.apa.org/doiLanding?doi=10.1037%2Fccp0000260. It's worth noting that research in the first decade of this century by the highly esteemed Peter Gollwitzer and his colleagues found that setting goals publicly (by sharing them with another person) reduced the likelihood that the sharer would achieve

those goals. Although it was a robust finding at the time, subsequent research has found that when people share their goals with someone else, they're more likely to attain those goals. I'm not sure how to resolve the discrepancy. Perhaps it's a cultural issue, as the benefits of private goal setting were found largely in Germany, or perhaps it's a generational issue that's been shaped by social media. Social media, which started in the early 2000s, may have made people more comfortable with and motivated by public goal-sharing. For the early research on how private goal-setting is better, see Peter M. Gollwitzer et al., "When Intentions Go Public: Does Social Reality Widen the Intention-Behavior Gap?," *Psychological Science* 20, no. 5 (2009): 612-618.

11. Teresa Amabile and Steven Kramer, *The Progress Principle: Using Small Wins to Ignite Joy, Engagement, and Creativity at Work* (Boston: Harvard Business Review Press, 2011).

12. Fiona H. McKay et al., "Using Health and Well-Being Apps for Behavior Change: A Systematic Search and Rating of Apps," *JMIR mHealth and uHealth* 7, no. 7 (2019): e11926, https://www.proquest.com/docview/2511246196 /fulltextPDF/321EFD90EFB043FFPQ/1?accountid=28598.

13. Sandra Wittleder et al., "Mental Contrasting with Implementation Intentions Reduces Drinking When Drinking Is Hazardous: An Online Self-Regulation Intervention," *Health Education & Behavior* 46, no. 4 (2019): 666-676, https://journals.sagepub.com/doi/full/10.1177/1090198119826284.

14. Peter M. Gollwitzer and Paschal Sheeran, "Implementation Intentions and Goal Achievement: A Meta-analysis of Effects and Processes," *Advances in Experimental Social Psychology* 38 (2006): 69-119, https://www.researchgate.net /profile/Peter-Gollwitzer 2/publication/37367696_Implementation _Intentions_and_Goal_Achievement_A_Meta-Analysis_of_Effects_and _Processes/links/59d91a24a6fdcc2aad0d8c1f/Implementation-Intentions -and-Goal-Achievement-A-Meta-Analysis-of-Effects-and-Processes.pdf?_sg%5 B0%5D=started_experiment_milestone&origin=journalDetail.

15. Benjamin Harkin et al., "Does Monitoring Goal Progress Promote Goal Attainment? A Meta-analysis of the Experimental Evidence," *Psychological Bulletin* 142, no. 2 (2016): 198, https://eprints.whiterose.ac.uk/91437/8/3_PDFsam _Does%20monitoring%20goal.pdf.

16. Minjung Koo and Ayelet Fishbach, "The Small-Area Hypothesis: Effects of Progress Monitoring on Goal Adherence," *Journal of Consumer Research* 39, no. 3 (2012): 493-509, https://www.jstor.org/stable/pdf/10.1086/663827.pdf ?casa_token=M2juUmGCS5oAAAAA:-lJUd4SCH8CKWiPnJHzYqQUYE_D -6jWugJ8Q6bD5TdjhnkxvOFLwaph2XXgKLwWcsit8oXE5fqRLsJqbGx8Y AjoCHJF5w_5RjzJ08r4QUQ3m0m_6Xi0.

Chapter 5

1. Sam J. Gilbert and Paul W. Burgess, "Executive Function," *Current Biology* 18, no. 3 (2008): R110-R114, https://www.cell.com/current-biology/pdf/S0960-9822(07)02367-6.pdf.

2. Grant S. Shields, Matthew A. Sazma, and Andrew P. Yonelinas, "The Effects of Acute Stress on Core Executive Functions: A Meta-analysis and Comparison with Cortisol," *Neuroscience & Biobehavioral Reviews* 68 (2016): 651-668, https://www.ncbi.nlm.nih.gov/pmc/articles/PMC5003767/.

3. Ibid.

4. Bassam Khoury et al., "Mindfulness-Based Stress Reduction for Healthy Individuals: A Meta-analysis," *Journal of Psychosomatic Research* 78, no. 6 (2015): 519-528, https://r.jordan.im/download/mindfulness/khoury2015.pdf.

5. Amit Mohan, Ratna Sharma, and Ramesh L. Bijlani, "Effect of Meditation on Stress-Induced Changes in Cognitive Functions," *Journal of Alternative and Complementary Medicine* 17, no. 3 (2011): 207-212, https://www.liebertpub.com/doi/abs/10.1089/acm.2010.0142.

6. Julia C. Basso and Wendy A. Suzuki, "The Effects of Acute Exercise on Mood, Cognition, Neurophysiology, and Neurochemical Pathways: A Review," *Brain Plasticity* 2, no. 2 (2017): 127-152.

7. "Target Heart Rates Chart," American Heart Association, last reviewed March 9, 2021, https://www.heart.org/en/healthy-living/fitness/fitness-basics/target-heart-rates.

8. "Exercise Intensity: How to Measure It," Mayo Clinic, August 25, 2023, https://www.mayoclinic.org/healthy-lifestyle/fitness/in-depth/exercise-intensity/art-20046887.

9. David Moreau and Edward Chou, "The Acute Effect of High-Intensity Exercise on Executive Function: A Meta-analysis," *Perspectives on Psychological Science* 14, no. 5 (2019): 734-764.

10. Julia C. Basso et al., "Acute Exercise Improves Prefrontal Cortex but Not Hippocampal Function in Healthy Adults," *Journal of the International Neuropsychological Society* 21, no. 10 (2015): 791-801.

11. Moreau and Chou, "The Effect of Exercise."

12. P. Zimmer et al., "The Effects of Different Aerobic Exercise Intensities on Serum Serotonin Concentrations and Their Association with Stroop Task Performance: A Randomized Controlled Trial," *European Journal of Applied Physiology* 116, no. 10 (2016): 2025-2034.

13. R. Pagliari and L. Peyrin, "Norepinephrine Release in the Rat Frontal Cortex Under Treadmill Exercise: A Study with Microdialysis," *Journal of Applied Physiology* 78, no. 6 (1995): 2121–2130.

14. Sheree F. Logue and Thomas J. Gould, "The Neural and Genetic Basis of Executive Function: Attention, Cognitive Flexibility, and Response Inhibition," *Pharmacology Biochemistry and Behavior* 123 (2014): 45-54.

15. Juan Arturo Ballester-Ferrer et al., "Effect of Acute Exercise Intensity on Cognitive Inhibition and Well-Being: Role of Lactate and BDNF Polymorphism in the Dose-Response Relationship," *Frontiers in Psychology* 13 (2022): 7635.

16. Tetsuo Ohkuwa et al., "The Relationship Between Exercise Intensity and Lactate Concentration on the Skin Surface," *International Journal of Biomedical Science* 5, no. 1 (2009): 23.

17. Ibid.

18. S. Ludyga et al., "Acute Effects of Moderate Aerobic Exercise on Specific Aspects of Executive Function in Different Age and Fitness Groups: A Meta-analysis," *Psychophysiology* 53, no. 11 (2016): 1611-1626.

19. F. T. Chen et al., "Effects of Exercise Training Interventions on Executive Function in Older Adults: A Systematic Review and Meta-analysis," *Sports Medicine* 50, no. 8 (2020): 1451-1467.

20. Adele Diamond and Daphne S. Ling, "Review of the Evidence on, and Fundamental Questions About, Efforts to Improve Executive Functions, Including Working Memory," *Cognitive and Working Memory Training: Perspectives from Psychology, Neuroscience, and Human Development* 143 (2019), https://www.researchgate.net/profile/Adele-Diamond/publication/337745861 _Review_of_the_Evidence_on_and_Fundamental_Questions_About_Efforts_to_Improve_Executive_Functions_Including_Working_Memory/links /5e693268a6fdcc759502f1e0/Review-of-the-Evidence-on-and-Fundamental -Questions-About-Efforts-to-Improve-Executive-Functions-Including-Working -Memory.pdf.

21. Kimberley Luu and Peter A. Hall, "Examining the Acute Effects of Hatha Yoga and Mindfulness Meditation on Executive Function and Mood," *Mindfulness* 8, no. 4 (2017): 873-880.

Chapter 6

1. Hats off to Gordy Slack, a science writer at Stanford University, who deserves credit for the librarian metaphor for the hippocampus. I first saw him use it in a blog post about the movie *Memento*: Gordy Slack, "What's That Movie Called? Memory in Film," *Brainstorm* (blog), August 19, 2010, https://gordyslack .blogspot.com/2010/08/forgetting-memento.html. Like all analogies, however, it has its limitations. Memories aren't really like "books shelved in a library" because books don't change over time. Sure, the binding and pages may deteriorate a little, maybe someone makes some notes in the margins, but the text and pictures in the books don't become new text and pictures. But memories do change

over time. Scientists currently hypothesize that when we retrieve a memory, we're actually reconstructing the memory, which means that it's not necessarily exactly the same as when we stored it. If you've ever had a family dinner where everyone at the table remembers the same event differently, it's not just that you had different perspectives—all of you are reconstructing the memories differently. You might be convinced that your memory is the correct one because it feels like you're replaying the video of what happened, but the "video" you've stored has probably been edited and you don't even realize it.

2. For the original study in all its glorious detail, see Nakul Yadav et al., "Prefrontal Feature Representations Drive Memory Recall," *Nature* 608, no. 7921 (2022): 153-160, https://www.nature.com/articles/s41586-022-04936-2. For a more user-friendly description of the research, see https://www.rockefeller.edu/news/32444-memory-fragments-stored-off-hippocampus-in-prefrontal-cortex/.

3. Mahmoud A. Alomari et al., "Forced and Voluntary Exercises Equally Improve Spatial Learning and Memory and Hippocampal BDNF Levels," *Behavioural Brain Research* 247 (2013): 34-39.

4. Cristen Brownlee, "Buff and Brainy," *Science News Online* 169, no. 8 (2006), http://www.entelia.com/members/Foundation%2011/F11.2%20www/Unit2%20Needing%20-%20Basic%20Reading/Food%20-%20Butt%20and%20Brainy.pdf.

5. For some thorough reading on what BDNF is, how exercise increases it, and how it doesn't cross the blood-brain barrier, see Gary Wenk's book, *Your Brain on Exercise* (Oxford: Oxford University Press, 2021).

6. N. Feter et al., "How Do Different Physical Exercise Parameters Modulate Brain-Derived Neurotrophic Factor in Healthy and Non-healthy Adults? A Systematic Review, Meta-analysis and Meta-regression," *Science & Sports* 34, no. 5 (2019): 293-304.

7. For research on how higher levels of BDNF protect against Alzheimer's, see Michal Schnaider Beeri and Joshua Sonnen, "Brain BDNF Expression as a Biomarker for Cognitive Reserve Against Alzheimer Disease Progression," *Neurology* 86, no. 8 (2016): 702-703. For research on treating Alzheimer's disease with BDNF, see Suzanne Gascon et al., "Peptides Derived from Growth Factors to Treat Alzheimer's Disease," *International Journal of Molecular Sciences* 22, no. 11 (2021): 6071. For research on the role of BDNF in depression and how it might be used to treat depression, see Tao Yang et al., "The Role of BDNF on Neural Plasticity in Depression," *Frontiers in Cellular Neuroscience* 14 (2020): 82, https://www.frontiersin.org/articles/10.3389/fncel.2020.00082/full.

8. Preeyam K. Parikh et al., "The Impact of Memory Change on Daily Life in Normal Aging and Mild Cognitive Impairment," *The Gerontologist* 56, no. 5 (2016): 877-885, https://academic.oup.com/gerontologist/article/56/5/877/2605267.

9. Xiangfei Meng et al., "Effects of Dance Intervention on Global Cognition, Executive Function and Memory of Older Adults: A Meta-analysis and Systematic

Review," *Aging Clinical and Experimental Research* 32 (2020): 7-19.

10. Marcelo de Maio Nascimento, "Dance, Aging, and Neuroplasticity: An Integrative Review," *Neurocase* 27, no. 4 (2021): 372-381.

11. Ibid. See also Agnieszka Z. Burzynska et al., "White Matter Integrity Declined over 6-Months, but Dance Intervention Improved Integrity of the Fornix of Older Adults," *Frontiers in Aging Neuroscience* 9 (2017), https://doi.org/10.3389/fnagi.2017.00059.

12. C.-J. Olsson, "Dancing Combines the Essence for Successful Aging," *Frontiers in Neuroscience* 6 (2012): 155.

13. Molly A. Youngs et al., "Mindfulness Meditation Improves Visual Short-Term Memory," *Psychological Reports* 124, no. 4 (2021): 1673-1686, https://journals.sagepub.com/doi/pdf/10.1177/0033294120926670.

14. Kirk Warren Brown et al., "Mindfulness Enhances Episodic Memory Performance: Evidence from a Multimethod Investigation," *PLoS One* 11, no. 4 (2016): e0153309, https://journals.plos.org/plosone/article?id=10.1371/journal.pone.0153309. The 75% improvement in recall is seen in Study 3, where the average number of items recalled was 4.32 for the control group and 7.58 for the meditation group. Not all studies find that mindfulness meditation improves memory, however, and researchers think the variance is due to the different kinds of mindfulness meditation techniques, which can vary widely, and some techniques may strengthen memory more than others.

15. Julia C. Basso et al., "Brief, Daily Meditation Enhances Attention, Memory, Mood, and Emotional Regulation in Non-Experienced Meditators," *Behavioural Brain Research* 356 (2019): 208-220, https://www.sciencedirect.com/science/article/pii/S016643281830322X?casa_token=_GT4Remhti8AAAAA:V74c0PzdiufgylXKNPxq2Y86qamvhhGqx_6Q9j4htvdUb8OrcL0GhRWzeugNOHnczOtvCxnB-g.

16. Ibid.

17. Susanna Feruglio et al., "The Impact of Mindfulness Meditation on the Wandering Mind: A Systematic Review," *Neuroscience & Biobehavioral Reviews* 131 (2021): 313-330, https://www.sciencedirect.com/science/article/abs/pii/S0149763421004140?casa_token=yrXIpFhzV48AAAAA:62RR6WXE5TEfoSRDS5plfRXrWOaiPDJ3MgYRyjAICIU47rObqf_YdKFMdiiTsLMHFwnt2rkDXQ.

18. Kathleen A. Garrison et al., "Meditation Leads to Reduced Default Mode Network Activity Beyond an Active Task," *Cognitive, Affective, & Behavioral Neuroscience* 15 (2015): 712-720, https://link.springer.com/content/pdf/10.3758/s13415-015-0358-3.pdf.

19. Marco Sperduti, Pénélope Martinelli, and Pascale Piolino, "A Neurocognitive Model of Meditation Based on Activation Likelihood Estimation (ALE) Meta-analysis," *Consciousness and Cognition* 21, no. 1 (2012): 269-276, https://www.sciencedirect.com/science/article/pii/S1053810011002285.

20. Ibid.

21. Britta K. Hölzel et al., "Mindfulness Practice Leads to Increases in Regional Brain Gray Matter Density," *Psychiatry Research: Neuroimaging* 191, no. 1 (2011): 36-43, https://www.sciencedirect.com/science/article/pii /S092549271000288X?casa_token=OKV0iPqzaS8AAAAA:6MCOCJFZ TI0L4tH-vk0d2s8mdGMCUcW4nWHTRsMS6WEHTBnUuYnZc5G7pasToK-g0j4vrIQrNNw. For a general overview of the many changes that happen to the brain as a result of mindfulness meditation, see Yi-Yuan Tang, Britta K. Hölzel, and Michael I. Posner, "The Neuroscience of Mindfulness Meditation," *Nature Reviews Neuroscience* 16, no. 4 (2015): 213-225, https://www.researchgate .net/profile/Britta-Holzel/publication/273774412_The_neuroscience_of _mindfulness_meditation/links/550ca4970cf27526109679f3/The-neuroscience -of-mindfulness-meditation.pdf.

22. Eileen Luders et al., "Meditation Effects Within the Hippocampal Complex Revealed by Voxel-Based Morphometry and Cytoarchitectonic Probabilistic Mapping," *Frontiers in Psychology* 4 (2013): 398, https://www.frontiersin.org /articles/10.3389/fpsyg.2013.00398/full.

23. Ruth Peters, "Ageing and the Brain," *Postgraduate Medical Journal* 82, no. 964 (2006): 84-88, https://www.ncbi.nlm.nih.gov/pmc/articles/PMC2596698 /#:~:text=As%20we%20age%20our%20brains,more%20bilateral%20for%20 memory%20tasks.

24. Kevin W. Chen et al., "Meditative Therapies for Reducing Anxiety: A Systematic Review and Meta-analysis of Randomized Controlled Trials," *Depression and Anxiety* 29, no. 7 (2012): 545-562, https://www.ncbi.nlm.nih.gov/pmc/articles /PMC3718554/?_escaped_fragment_=po=6.57895.

25. Getu Gamo Sagaro, Enea Traini, and Francesco Amenta, "Activity of Choline Alphoscerate on Adult-Onset Cognitive Dysfunctions: A Systematic Review and Meta-analysis," *Journal of Alzheimer's Disease* 92, no. 1 (2023): 59-70, https://content.iospress.com/articles/journal-of-alzheimers-disease/jad221189.

26. Adam G. Parker et al., "The Effects of Alpha-glycerylphosphorylcholine, Caffeine or Placebo on Markers of Mood, Cognitive Function, Power, Speed, and Agility," *Journal of the International Society of Sports Nutrition* 12, suppl.[1] (2015): 41.

27. The classic study that's cited on the benefits of testing was done by the great memory researcher, Henry Roediger III. See Henry L. Roediger III and Jeffrey D. Karpicke, "Test-Enhanced Learning: Taking Memory Tests Improves Long-Term Retention," *Psychological Science* 17, no. 3 (2006): 249-255, https://journals .sagepub.com/doi/abs/10.1111/j.1467-9280.2006.01693.x?url_ver=Z39.88 -2003&rfr_id=ori:rid:crossref.org&rfr_dat=cr_pub%3dpubmed. More recent analyses reveal that testing yourself remains one of the best ways to hold onto information that you don't want to forget: See Chunliang Yang et al., "Testing (Quizzing) Boosts Classroom Learning: A Systematic and Meta-analytic Review," *Psychological Bulletin* 147, no. 4 (2021): 399, https://pubmed.ncbi.nlm.nih.gov

/33683913/.

28. Dorottya Bencze et al., "An Event-Related Potential Study of the Testing Effect: Electrophysiological Evidence for Context-Dependent Processes Changing Throughout Repeated Practice," *Biological Psychology* 171 (2022): 108341, https://www.sciencedirect.com/science/article/pii/S0301051122000837.

29. Gesa Van den Broek et al., "Neurocognitive Mechanisms of the 'Testing Effect': A Review," *Trends in Neuroscience and Education* 5, no. 2 (2016): 52-66, https://pure.mpg.de/rest/items/item_2300773_4/component/file_2351433 /content.

30. Serge Brédart, "Strategies to Improve Name Learning," *European Psychologist* (2019), https://orbi.uliege.be/bitstream/2268/238248/1/NameLearning2019 EPVersionOCR.pdf.

31. Christa K. McIntyre and Benno Roozendaal, "Adrenal Stress Hormones and Enhanced Memory for Emotionally Arousing Experiences," in *Neural Plasticity and Memory: From Genes to Brain Imaging*, ed. Federico Bermúdez-Rattoni (Boca Raton, FL: CRC Press/Taylor & Francis, 2007), 265, https://www.ncbi .nlm.nih.gov/books/NBK3907/.

32. Ibid. See also Larry Cahill, Lukasz Gorski, and Kathryn Le, "Enhanced Human Memory Consolidation with Post-learning Stress: Interaction with the Degree of Arousal at Encoding," *Learning & Memory* 10, no. 4 (2003): 270-274, https://www.ncbi.nlm.nih.gov/pmc/articles/PMC202317/.

Chapter 7

1. Carol S. Dweck, *Mindset: The New Psychology of Success* (New York: Random House, 2006).

2. Ibid., p. 33.

3. David S. Yeager and Carol S. Dweck, "What Can Be Learned from Growth Mindset Controversies?," *American Psychologist* 75, no. 9 (2020): 1269, https://psycnet.apa.org/manuscript/2020-99903-019.pdf.

4. Robert C. Wilson et al., "The Eighty Five Percent Rule for Optimal Learning," *Nature Communications* 10, no. 1 (2019): 4646, https://www.nature.com /articles/s41467-019-12552-4.

5. For a user-friendly version of the research, see https://news.yale.edu/2018/07 /19/arent-sure-brain-primed-learning. For the original scientific publication, see Bart Massi, Christopher H. Donahue, and Daeyeol Lee, "Volatility Facilitates Value Updating in the Prefrontal Cortex," *Neuron* 99, no. 3 (2018): 598-608, https://www.cell.com/neuron/fulltext/S0896-6273(18)30529-4.

6. Jason S. Moser et al., "Mind Your Errors: Evidence for a Neural Mechanism Linking Growth Mind-Set to Adaptive Posterror Adjustments," *Psychological Science* 22, no. 12 (2011): 1484-1489, https://cpl.psy.msu.edu/wp-content

/uploads/2020/01/Moser-et-al.-2011.pdf.

7. Ibid.

8. Jennifer A. Mangels et al., "Why Do Beliefs About Intelligence Influence Learning Success? A Social Cognitive Neuroscience Model," *Social Cognitive and Affective Neuroscience* 1, no. 2 (2006): 75-86, https://academic.oup.com/scan/article/1/2/75/2362769?view=extract.

9. Yeager and Dweck, "What Can Be Learned from Growth Mindset Controversies?"

10. Moser et al., "Mind Your Errors."

11. Yeager and Dweck, "What Can Be Learned from Growth Mindset Controversies?"

12. Vyara Valkanova, Rocio Eguia Rodriguez, and Klaus P. Ebmeier, "Mind over Matter—What Do We Know About Neuroplasticity in Adults?," *International Psychogeriatrics* 26, no. 6 (2014): 891-909.

13. Hye Rin Lee et al., "Components of Engagement in Saying-Is-Believing Exercises," *Current Psychology* (2022): 1-16, https://link.springer.com/article/10.1007/s12144-022-02782-z.

14. For a list of benefits associated with having a growth mindset at work, see Soo Jeoung Han and Vicki Stieha, "Growth Mindset for Human Resource Development: A Scoping Review of the Literature with Recommended Interventions," *Human Resource Development Review* 19, no. 3 (2020): 309-331, https://journals.sagepub.com/doi/pdf/10.1177/1534484320939739?casa_token=ItXohqk6kZMAAAAA:fTacb09dOBRQF_fe3ZKwJnG56PJE3JPMi9p-NSccw0KJhZl_TW30vfcJm1E2lpJQn95r_NiX5Qnk.

15. Aneeta Rattan and Carol S. Dweck, "Who Confronts Prejudice? The Role of Implicit Theories in the Motivation to Confront Prejudice," *Psychological Science* 21, no. 7 (2010): 952-959, https://journals.sagepub.com/doi/full/10.1177/0956797610374740?casa_token=ZVWUGy5u2VkAAAAA:bSI6W DtmYd-wF8dmrBczalIJo0UohtR7ffbUb08SeoHd08De8QGJENKjZzyd376n DJ7aFqR4aSzS.

Chapter 8

1. Benjamin M. P. Cuff et al., "Empathy: A Review of the Concept," *Emotion Review* 8, no. 2 (2016): 144-153, https://journals.sagepub.com/doi/full/10.1177/1754073914558466?casa_token=ZcLWtWrztEoAAAAA:V28BFCQqmd6s_p2fUTxkQK6-sFJXg03q9ec8T9KoFbIW6QcUW9x__NJT9ZtPplHyPN-pCrcOChll.

2. Julia Stietz et al., "Dissociating Empathy from Perspective-Taking: Evidence from Intra- and Inter-individual Differences Research," *Frontiers in Psychiatry* 10 (2019): 126, https://www.frontiersin.org/journals/psychiatry/articles/10.3389/fpsyt.2019.00126/full. See also Christine L. Cox et al., "The Balance Between Feeling and Knowing: Affective and Cognitive Empathy Are Reflected in the

Brain's Intrinsic Functional Dynamics," *Social Cognitive and Affective Neuroscience* 7, no. 6 (2012): 727-737, https://academic.oup.com/scan/article/7/6/727/1645655?login=false.

3. Tara Van Bommel, "The Power of Empathy in Times of Crisis and Beyond," Catalyst (2021), https://www.catalyst.org/reports/empathy-work-strategy-crisis.

4. Tracy Brower, "Empathy Is the Most Important Leadership Skill According to Research," *Forbes*, September 19, 2021, https://www.forbes.com/sites/tracy brower/2021/09/19/empathy-is-the-most-important-leadership-skill-according-to-research/?sh=68aade243dc5.

5. Jamil Zaki, "Empathy: A Motivated Account," *Psychological Bulletin* 140, no. 6 (2014): 1608, https://static1.squarespace.com/static/55917f64e4b0cd3b4705b68c/t/5c732030e4966b4ed5335902/1551048756386/zaki.2014.pdf.

6. Erika Weisz and Jamil Zaki, "Empathy Building Interventions: A Review of Existing Work and Suggestions for Future Directions," In *The Oxford Handbook of Compassion Science*, eds. E. M. Seppälä et al. (Oxford: Oxford University Press, 2017), 205-217.

7. Ibid. For a more recent review of how motives affect empathy, see Erika Weisz and Jamil Zaki, "Motivated Empathy: A Social Neuroscience Perspective," *Current Opinion in Psychology* 24 (2018): 67-71, https://www.sciencedirect.com/science/article/pii/S2352250X18300150?casa_token=5sEaZ-K6x7EAAAAA:bMWzydhCzpetukZE7IHpccLg_5jx7n6ZzwCu5p7BfNV02kw3zrm4Zq2QROPCIEVImudocR0oUA.

8. C. Daryl Cameron, Lasana T. Harris, and B. Keith Payne, "The Emotional Cost of Humanity: Anticipated Exhaustion Motivates Dehumanization of Stigmatized Targets," *Social Psychological and Personality Science* 7, no. 2 (2016): 105-112, https://journals.sagepub.com/doi/10.1177/1948550615604453.

9. María José Gutiérrez-Cobo et al., "Does Our Cognitive Empathy Diminish with Age? The Moderator Role of Educational Level," *International Psychogeriatrics* 35, no. 4 (2023): 207-214, https://pubmed.ncbi.nlm.nih.gov/34078514/.

10. Ibid.

11. Binghai Sun et al., "Lack of Interaction Motivation in Older Adults Automatically Reduces Cognitive Empathy," *Experimental Aging Research* 50, no. 2 (2024): 225-247, https://www.tandfonline.com/doi/abs/10.1080/0361073X.2023.2168990.

12. Amy L. Jarvis et al., "Emotional Empathy Across Adulthood: A Meta-analytic Review," *Psychology and Aging* 39, no. 2 (2023): 126-138, https://psycnet.apa.org/record/2024-26378-001.

13. Katja Wiech et al., "Anterior Insula Integrates Information About Salience into Perceptual Decisions About Pain," *Journal of Neuroscience* 30, no. 48 (2010): 16324-16331, https://www.jneurosci.org/content/jneuro/30/48/16324.full.pdf.

14. Corrado Corradi-Dell'Acqua et al., "Cross-Modal Representations of First-Hand and Vicarious Pain, Disgust and Fairness in Insular and Cingulate Cortex," *Nature Communications* 7, no. 1 (2016): 10904, https://www.nature.com/articles/ncomms10904.

15. Gabriele Chierchia and Tania Singer, "The Neuroscience of Compassion and Empathy and Their Link to Prosocial Motivation and Behavior," in *Decision Neuroscience,* eds. Jean-Claude Dreher and Léon Tremblay (Cambridge, MA: Academic Press, 2017), 247-257, https://www.sciencedirect.com/science/article/abs/pii/B9780128053089000208.

16. C. Daryl Cameron et al., "Empathy Is Hard Work: People Choose to Avoid Empathy Because of Its Cognitive Costs," *Journal of Experimental Psychology: General* 148, no. 6 (2019): 962, https://psycnet.apa.org/manuscript/2019-20830-001.pdf.

17. Francis Stevens and Katherine Taber, "The Neuroscience of Empathy and Compassion in Pro-social Behavior," *Neuropsychologia* 159, no. 107925 (2021): 6, http://change-et-sois.org/wp-content/uploads/2023/01/The-neuroscience-of-empathy-and-compassion-in-pro-social-behavior-Stevens-F-Taber-K-2021.pdf.

18. Olga M. Klimecki et al., "Differential Pattern of Functional Brain Plasticity After Compassion and Empathy Training," *Social Cognitive and Affective Neuroscience* 9, no. 6 (2014): 873-879, https://academic.oup.com/scan/article/9/6/873/1669505?login=false.

19. Specifically, the nucleus accumbens, a small part of the ventral striatum, is activated during charitable giving and feelings of strong positive affect for the person to whom you're giving. See Alexander Genevsky et al., "Neural Underpinnings of the Identifiable Victim Effect: Affect Shifts Preferences for Giving," *Journal of Neuroscience* 33, no. 43 (2013): 17188-17196, https://www.jneurosci.org/content/jneuro/33/43/17188.full.pdf. See also Alexander Genevsky and Brian Knutson, "Neural Affective Mechanisms Predict Market-Level Microlending," *Psychological Science* 26, no. 9 (2015): 1411-1422, https://journals.sagepub.com/doi/full/10.1177/0956797615588467.

20. For a review of studies that found that compassion training improves helping (or what's known in psychology as "prosocial behavior"), see Olga M. Klimecki and Tania Singer, "The Compassionate Brain," in Seppälä et al., *The Oxford Handbook of Compassion Science,* 109-120, https://www.unige.ch/fapse/e3lab/files/8215/4072/9893/Klimecki_2017_BC.pdf.

21. For research on experiencing more positive emotions with regular self-compassion practice, see Kristin Neff and Christopher Germer, "Self-Compassion and Psychological Well-Being," in Seppälä et al., *The Oxford Handbook of Compassion Science,* 371-390. For research on loving-kindness meditation and vagal tone, see Bethany E. Kok et al., "How Positive Emotions Build Physical Health: Perceived Positive Social Connections Account for the Upward Spiral Between Positive Emotions and Vagal Tone," *Psychological Science* 24, no. 7

(2013): 1123-1132, https://www.mentorcoach.com/wp-content/uploads/2017/05/Week-3-Reading-Kok-et-al-2013.pdf. For research on longer telomeres with loving-kindness meditation, see Khoa D. Le Nguyen et al., "Loving-Kindness Meditation Slows Biological Aging in Novices: Evidence from a 12-Week Randomized Controlled Trial," *Psychoneuroendocrinology* 108 (2019): 20-27, https://www.sciencedirect.com/science/article/abs/pii/S0306453019300010.

22. For data on more compassionate care for patients, see Shane Sinclair et al., "What Works for Whom in Compassion Training Programs Offered to Practicing Healthcare Providers: A Realist Review," *BMC Medical Education* 21, no. 1 (2021): 455, https://link.springer.com/content/pdf/10.1186/s12909-021-02863-w.pdf. For data on reduced burnout and emotional exhaustion, see Ian M. Kratzke, "Self-Compassion Training to Improve Well-Being for Surgical Residents," *EXPLORE* 19, no. 1 (2023): 78-83, https://www.sciencedirect.com/science/article/abs/pii/S1550830722000714. For data on reduced interpersonal conflict, see Ciro Conversano et al., "Mindfulness, Compassion, and Self-Compassion Among Health Care Professionals: What's New? A Systematic Review," *Frontiers in Psychology* 11 (2020): 1683, https://www.frontiersin.org/articles/10.3389/fpsyg.2020.01683/full.

23. Jason Mills and Michael Chapman, "Compassion and Self-Compassion in Medicine: Self-Care for the Caregiver," *Australasian Medical Journal* 9, no. 5 (2016): 87-91, https://eprints.qut.edu.au/107931/1/2583-13266-1-PB-1.pdf.

Chapter 9

1. The classic paper on this topic is Marilynn B. Brewer, "The Psychology of Prejudice: Ingroup Love and Outgroup Hate?," *Journal of Social Issues* 55, no. 3 (1999): 429-444, http://courses.washington.edu/pbafhall/563/Readings/Brewer.pdf. For a more recent meta-analysis affirming that ingroup favoritism continues to be stronger than outgroup derogation, see Daniel Balliet, Junhui Wu, and Carsten K. W. De Dreu, "Ingroup Favoritism in Cooperation: A Meta-analysis," *Psychological Bulletin* 140, no. 6 (2014): 1556, https://amsterdamcooperationlab.com/wp-content/uploads/2015/11/balliet-wu-de-dreu_advanced-online-2014.pdf.

2. Tiffany A. Ito and Geoffrey R. Urland, "Race and Gender on the Brain: Electrocortical Measures of Attention to the Race and Gender of Multiply Categorizable Individuals," *Journal of Personality and Social Psychology* 85, no. 4 (2003): 616, https://web.archive.org/web/20170829185112id_/http://psych.colorado.edu/~tito/Ito&Urland2003.pdf.

3. Eric J. Vanman, "The Role of Empathy in Intergroup Relations," *Current Opinion in Psychology* 11 (2016): 59-63, https://psycnet.apa.org/record/2016-48343-016. Or see Shihui Han, "Neurocognitive Basis of Racial Ingroup Bias in Empathy," *Trends in Cognitive Sciences* 22, no. 5 (2018): 400-421, http://www

.psy.pku.edu.cn/docs/20181226142849692633.pdf.

4. Tiffani J. Johnson et al., "Association of Race and Ethnicity with Management of Abdominal Pain in the Emergency Department," *Pediatrics* 132, no. 4 (2013): e851-e858, https://publications.aap.org/pediatrics/article-abstract/132/4/e851 /64891/Association-of-Race-and-Ethnicity-With-Management?redirectedFrom =fulltext.

5. Wenxin Li and Shihui Han, "Behavioral and Electctrophysiological Evidence for Enhanced Sensitivity to Subtle Variations of Pain Expressions of Same-Race Than Other-Race Faces," *Neuropsychologia* 129 (2019): 302-309, https://www .sciencedirect.com/science/article/pii/S0028393219300818?casa_token =nPvOSlpBj4MAAAAA:NMKNic_HUuXibx5YWzQDrLQSH8sVOaq6dNL HyHFzphWu2bJy1sQez2dPDws1W-fs8zrLZuR_Ag.

6. Eva H. Telzer, Nicolas Ichien, and Yang Qu, "The Ties That Bind: Group Membership Shapes the Neural Correlates of In-Group Favoritism," *NeuroImage* 115 (2015): 42-51, https://bpb-us-e1.wpmucdn.com/sites.northwestern.edu/dist /7/6360/files/2022/03/Telzer-Ichien-Qu-2015-NeuroImage-In-group-favoritism .pdf.

7. Matt T. Richins et al., "Incidental Fear Reduces Empathy for an Out-Group's Pain," *Emotion* 21, no. 3 (2021): 536, https://ore.exeter.ac.uk/repository /bitstream/handle/10871/40355/Richins_Barreto_Fear_Intergroup_Empathy %20Emotion%202019.pdf?sequence=1.

8. Loren J. Martin et al., "Reducing Social Stress Elicits Emotional Contagion of Pain in Mouse and Human Strangers," *Current Biology* 25, no. 3 (2015): 326-332, https://www.cell.com/current-biology/pdf/S0960-9822(14)01489-4.pdf.

9. To identify strategies that are less effective at reducing bias and prejudice, I turned to a meta-analysis that averaged the results from many studies to sort the most effective strategies from the least effective. If you're a statistics geek like me, you might want to know that the two strategies I've listed here have effect sizes ranging from .10 to .28, whereas for the strategies I've listed that are most effective, the effect sizes are much higher, .37 to .43. See Elizabeth Levy Paluck et al., "Prejudice Reduction: Progress and Challenges," *Annual Review of Psychology* 72 (2021): 533-560, https://www.annualreviews.org/doi/full/10.1146/annurev -psych-071620-030619.

10. Inga K. Rösler and David M. Amodio, "Neural Basis of Prejudice and Prejudice Reduction," *Biological Psychiatry: Cognitive Neuroscience and Neuroimaging* 7, no. 12 (2022): 1200-1208, https://amodiolab.org/wp-content/uploads/2023/03 /Ro%CC%88sler-Amodio-2022.pdf.

11. David M. Amodio, Patricia G. Devine, and Eddie Harmon-Jones, "A Dynamic Model of Guilt: Implications for Motivation and Self-Regulation in the Context of Prejudice," *Psychological Science* 18, no. 6 (2007): 524-530, https://journals .sagepub.com/doi/10.1111/j.1467-9280.2007.01933.x.

12. Doyin Atewologun, Tinu Cornish, and Fatima Tresh, "Unconscious Bias: Train-

ing," *An Assessment of the Evidence for Effectiveness. Equality and Human Rights Commission Research Report* 113 (2018).

13. Toni Schmader, Tara C. Dennehy, and Andrew S. Baron, "Why Antibias Interventions (Need Not) Fail," *Perspectives on Psychological Science* 17, no. 5 (2022): 1381-1403, https://journals.sagepub.com/doi/pdf/10.1177/17456916211057565.

14. Natalie R. Hall, Richard J. Crisp, and Mein-woei Suen, "Reducing Implicit Prejudice by Blurring Intergroup Boundaries," *Basic and Applied Social Psychology* 31, no. 3 (2009): 244-254, https://www.tandfonline.com/doi/abs/10.1080/01973530903058474.

15. Sohad Murrar and Markus Brauer, "Entertainment-Education Effectively Reduces Prejudice," *Group Processes & Intergroup Relations* 21, no. 7 (2018): 1053-1077, https://journals.sagepub.com/doi/pdf/10.1177/1368430216682350?casa_token=YMhNvNVdxPUAAAAA:KnGmHbXpH0w0Is6zC1fb0YxC0EXoWsmowcZcswLKQQrIYKOxgcsbyOCQHC36jrAQ8vr4HLPL54-X.

16. Ibid.

17. Leor M. Hackel, Jamil Zaki, and Jay J. Van Bavel, "Social Identity Shapes Social Valuation: Evidence from Prosocial Behavior and Vicarious Reward," *Social Cognitive and Affective Neuroscience* 12, no. 8 (2017): 1219-1228, https://academic.oup.com/scan/article/12/8/1219/3574675.

Chapter 10

1. Amy Sinden, "Cost-Benefit Analysis, Ben Franklin, and the Supreme Court," *UC Irvine Law Review* 4, no. 4 (2014): 1175, https://scholarship.law.uci.edu/cgi/viewcontent.cgi?article=1178&context=ucilr.

2. These stories about Elliot are recounted in Antonio Damasio's classic book, *Descartes Error: Emotion, Reason, and the Human Brain* (New York: Random House, 2006).

3. For research on the relationship between stock market returns and the weather, see David Hirshleifer and Tyler Shumway, "Good Day Sunshine: Stock Returns and the Weather," *Journal of Finance* 58, no. 3 (2003): 1009-1032, https://onlinelibrary.wiley.com/doi/abs/10.1111/1540-6261.00556. For research on the relationship between the World Cup and stock market returns, see Alex Edmans, Diego Garcia, and Øyvind Norli, "Sports Sentiment and Stock Returns," *Journal of Finance* 62, no. 4 (2007): 1967-1998, https://leeds-faculty.colorado.edu/garcia/paper91v32.pdf.

4. This clever reframing to "What will I do if/when . . ." comes from a great Inc.com article: Sarah Peck, "4 Simple Strategies to Help You Make Decisions Faster," Inc.com, September 19, 2018, https://www.inc.com/sarah-peck/4-simple-strategies-to-help-you-make-decisions-faster.html.

5. Norris Krueger Jr. and Peter R. Dickson, "How Believing in Ourselves Increases Risk Taking: Perceived Self-Efficacy and Opportunity Recognition," *Decision Sciences* 25, no. 3 (1994): 385-400, https://onlinelibrary.wiley.com/doi/abs/10.1111/j.1540-5915.1994.tb00810.x.

6. Paul C. Nutt, "The Identification of Solution Ideas During Organizational Decision Making," *Management Science* 39, no. 9 (1993): 1071-1085, https://pubs online.informs.org/doi/abs/10.1287/mnsc.39.9.1071.

7. For the research on changing your time perspective to achieve more decision clarity, see J. Edward Russo and Paul J. H. Schoemaker, *Winning Decisions: Getting It Right the First Time* (New York: Currency, 2002).

8. Julian F. Thayer et al., "A Meta-analysis of Heart Rate Variability and Neuro-imaging Studies: Implications for Heart Rate Variability as a Marker of Stress and Health," *Neuroscience & Biobehavioral Reviews* 36, no. 2 (2012): 747-756, https://www.sciencedirect.com/science/article/abs/pii/S0149763411002077.

9. Coleman O. Martin et al., "The Effects of Vagus Nerve Stimulation on Decision-Making," *Cortex* 40, no. 4-5 (2004): 605-612, https://www.sciencedirect.com /science/article/abs/pii/S0010945208701564. See also Thayer et al. in the previous note for a meta-analysis, reviewing normal populations.

10. James Douglas Bremner et al., "Application of Noninvasive Vagal Nerve Stimulation to Stress-Related Psychiatric Disorders," *Journal of Personalized Medicine* 10, no. 3 (2020): 119, https://www.mdpi.com/2075-4426/10/3/119.

11. Patricia H. Janak and Kay M. Tye, "From Circuits to Behaviour in the Amygdala," *Nature* 517, no. 7534 (2015): 284-292, https://www.ncbi.nlm.nih.gov/pmc /articles/PMC4565157/. See also Anushka B. P. Fernando, Jennifer E. Murray, and Amy L. Milton, "The Amygdala: Securing Pleasure and Avoiding Pain," *Frontiers in Behavioral Neuroscience* 7 (2013): 190, https://www.proquest.com /openview/f9f6e66abd5433154b8650fa7b2880a8/1?pq-origsite=gscholar &cbl=2046456.

12. For research on invigorating people to work harder, see Monja P. Neuser et al., "Vagus Nerve Stimulation Boosts the Drive to Work for Rewards," *Nature Communications* 11, no. 1 (2020): 3555, https://www.nature.com/articles /s41467-020-17344-9. For research on waking people from comas, see Marie M. Vitello et al., "Transcutaneous Vagal Nerve Stimulation to Treat Disorders of Consciousness: Protocol for a Double-Blind Randomized Controlled Trial," *International Journal of Clinical and Health Psychology* 23, no. 2 (2023): 100360, https://www.sciencedirect.com/science/article/pii/S1697260022000680.

13. Longer exhales have been tied to improved vagal stimulation, as measured by heart rate variability, in several studies, including Ilse Van Diest et al., "Inhalation/Exhalation Ratio Modulates the Effect of Slow Breathing on Heart Rate Variability and Relaxation," *Applied Psychophysiology and Biofeedback* 39 (2014): 171-180, https://link.springer.com/article/10.1007/s10484-014-9253-x.

14. Marijke De Couck et al., "How Breathing Can Help You Make Better Decisions:

Two Studies on the Effects of Breathing Patterns on Heart Rate Variability and Decision-Making in Business Cases," *International Journal of Psychophysiology* 139 (2019): 1-9, https://www.sciencedirect.com/science/article/abs/pii/S0167876018303258.

15. For research on how all-cause mortality and cardiovascular disease go down with higher HRV, see Su-Chen Fang, Yu-Lin Wu, and Pei-Shan Tsai, "Heart Rate Variability and Risk of All-Cause Death and Cardiovascular Events in Patients with Cardiovascular Disease: A Meta-analysis of Cohort Studies," *Biological Research for Nursing* 22, no. 1 (2020): 45-56, https://journals.sagepub.com/doi/10.1177/1099800419877442. For research on how higher HRV is associated with lower anxiety, see John A. Chalmers et al., "Anxiety Disorders Are Associated with Reduced Heart Rate Variability: A Meta-analysis," *Frontiers in Psychiatry* 5 (2014): 80, https://www.frontiersin.org/articles/10.3389/fpsyt.2014.00080/full. For research on how higher HRV is associated with a lower risk of Alzheimer's disease and cognitive impairment, see Roberto Zulli et al., "QT Dispersion and Heart Rate Variability Abnormalities in Alzheimer's Disease and in Mild Cognitive Impairment," *Journal of the American Geriatrics Society* 53, no. 12 (2005): 2135-2139, https://agsjournals.onlinelibrary.wiley.com/doi/abs/10.1111/j.1532-5415.2005.00508.x.

16. Katrina Hinde, Graham White, and Nicola Armstrong, "Wearable Devices Suitable for Monitoring Twenty Four Hour Heart Rate Variability in Military Populations," *Sensors* 21, no. 4 (2021): 1061, https://www.ncbi.nlm.nih.gov/pmc/articles/PMC7913967/.

Chapter 11

1. Robert Jütte, "The Early History of the Placebo," *Complementary Therapies in Medicine* 21, no. 2 (2013): 94-97, https://europepmc.org/article/med/23497809.

2. For research on reducing pain and other symptoms, see the rest of this chapter. For research on reducing the necessary amounts of medication for treatment by 50%, see research on ADHD: Adrian D. Sandler et al., "Conditioned Placebo Dose Reduction: A New Teatment in ADHD," *Journal of Developmental and Behavioral Pediatrics* 31, no. 5 (2010): 369, https://www.ncbi.nlm.nih.gov/pmc/articles/PMC2902360/.

3. Fabrizio Benedetti, *Placebo Effects* (New York: Oxford University Press, 2020).

4. Matthias Zunhammer et al., "Meta-analysis of Neural Systems Underlying Placebo Analgesia from Individual Participant fMRI Data," *Nature Communications* 12, no. 1 (2021): 1391, https://www.nature.com/articles/s41467-021-21179-3.

5. Tor D. Wager, David J. Scott, and Jon-Kar Zubieta, "Placebo Effects on Human-Opioid Activity During Pain." *Proceedings of the National Academy of Sciences* 104, no. 26 (2007): 11056-11061, https://www.pnas.org/doi/full/10.1073/pnas

.0702413104.

6. Fabrizio Benedetti et al., "The Specific Effects of Prior Opioid Exposure on Placebo Analgesia and Placebo Respiratory Depression," *Pain* 75, no. 2-3 (1998): 313-319, https://www.sciencedirect.com/science/article/pii/S0304395998000104.

7. Harvard Health Publishing, "The Power of the Placebo Effect" (December 13, 2021), https://www.health.harvard.edu/mental-health/the-power-of-the-placebo-effect.

8. For relieving symptoms of the common cold, see David Rakel et al., "Perception of Empathy in the Therapeutic Encounter: Effects on the Common Cold," *Patient Education and Counseling* 85, no. 3 (2011): 390-397, https://www.ncbi.nlm.nih.gov/pmc/articles/PMC3107395/. For improved social function and communication among children with autism, see A. A. Masi et al., "Predictors of Placebo Response in Pharmacological and Dietary Supplement Treatment Trials in Pediatric Autism Spectrum Disorder: A Meta-analysis," *Translational Psychiatry* 5, no. 9 (2015): e640, https://www.nature.com/articles/tp2015143. For treating a variety of diseases, see Benedetti, *Placebo Effects*.

9. Kevin M. McKay, Zac E. Imel, and Bruce E. Wampold, "Psychiatrist Effects in the Psychopharmacological Treatment of Depression," *Journal of Affective Disorders* 92, no. 2-3 (2006): 287-290, https://www.sciencedirect.com/science/article/abs/pii/S0165032706000395.

10. Lauren C. Howe, J. Parker Goyer, and Alia J. Crum, "Harnessing the Placebo Effect: Exploring the Influence of Physician Characteristics on Placebo Response," *Health Psychology* 36, no. 11 (2017): 1074, https://psycnet.apa.org/manuscript/2017-10534-001.pdf.

11. For research on partners' levels of oxytocin and hugging, see Kathleen C. Light, Karen M. Grewen, and Janet A. Amico, "More Frequent Partner Hugs and Higher Oxytocin Levels Are Linked to Lower Blood Pressure and Heart Rate in Premenopausal Women," *Biological Psychology* 69, no. 1 (2005): 5-21, https://www.mzellner.com/page4/files/2005-light.pdf. For research on parents' levels of oxytocin and how they affect their interactions with their infants, see Ruth Feldman, Ilanit Gordon, and Orna Zagoory-Sharon, "Maternal and Paternal Plasma, Salivary, and Urinary Oxytocin and Parent–Infant Synchrony: Considering Stress and Affiliation Components of Human Bonding," *Developmental Science* 14, no. 4 (2011): 752-761, https://onlinelibrary.wiley.com/doi/abs/10.1111/j.1467-7687.2010.01021.x.

12. Linda Handlin et al., "Short-Term Interaction Between Dogs and Their Owners: Effects on Oxytocin, Cortisol, Insulin and Heart Rate—an Exploratory Study," *Anthrozoös* 24, no. 3 (2011): 301-315, https://thehealthsciencesacademy.org/wp-content/uploads/2014/12/Dogs.Owners.Oxytocin.pdf.

13. Bonnie Auyeung et al., "Oxytocin Increases Eye Contact During a Real-Time, Naturalistic Social Interaction in Males with and Without Autism," *Translational Psychiatry* 5, no. 2 (2015): e507, https://www.nature.com/articles/tp2014146.

14. Carolyn H. Declerck, Christophe Boone, and Toko Kiyonari, "Oxytocin and Co-operation Under Conditions of Uncertainty: The Modulating Role of Incentives and Social Information," *Hormones and Behavior* 57, no. 3 (2010): 368-374, https://www.sciencedirect.com/science/article/pii/S0018506X10000188?casa _token=B1MnMTBpSGIAAAAA:7Bp9I7sDteb9S4qj7oA8wxz7uQS2raaXk NWRA9SwHiXbbn9iOLmCuPeGvHYZTcdUh7cbyYx-PQ.

15. What's with all the nasal sprays, you're wondering? The nose is, not surprisingly, a pretty quick way to reach the brain. Researchers can deliver oxytocin through a needle injection or a nasal spray, but they prefer the spray because it's much less intrusive and more of the hormone reaches the brain. Pills lose their potency by the time they make it all the way through the digestive tract. For research on how oxytocin increases trust, see Thomas Baumgartner et al., "Oxytocin Shapes the Neural Circuitry of Trust and Trust Adaptation in Humans," *Neuron* 58, no. 4 (2008): 639-650, https://www.cell.com/neuron/pdf/S0896-6273(08)00327-9 .pdf. For an explanation of why sprays are the preferred way to go for oxytocin delivery, see Adam J. Guatella et al., "Recommendations for the Standardisation of Oxytocin Nasal Administration and Guidelines for Its Reporting in Human Research," *Psychoneuroendocrinology* 38, no. 5 (2013): 612-625, https://www .sciencedirect.com/science/article/pii/S0306453012004118?casa_token=4S2Qlh _z4x4AAAAA:Dp26WE20FTgjHLM4zuUo9MvnIcZX3HS-pqFOyH3IQAkq0 AGpMhXV4TEnf-IG2hwgAh06UsuOdQ.

16. Elena Itskovich et al., "Oxytocin and the Social Facilitation of Placebo Effects," *Molecular Psychiatry* 27, no. 6 (2022): 2640-2649, https://www.ncbi.nlm.nih .gov/pmc/articles/PMC9167259/.

17. Howe et al., "Harnessing the Placebo Effect."

18. Marie P. Cross et al., "How and Why Could Smiling Influence Physical Health? A Conceptual Review," *Health Psychology Review* 17, no. 2 (2023): 321-343, https://www.tandfonline.com/doi/abs/10.1080/17437199.2022.2052740.

Chapter 12

1. Christian E. Waugh, Elaine Z. Shing, and R. Michael Furr, "Not All Disengage-ment Coping Strategies Are Created Equal: Positive Distraction, but Not Avoid-ance, Can Be an Adaptive Coping Strategy for Chronic Life Stressors," *Anxiety, Stress, and Coping* 33, no. 5 (2020): 511-529, https://www.tandfonline.com/doi /abs/10.1080/10615806.2020.1755820.

2. These items are adapted from Waugh, Shing, and Furr, "Not All Disengagement Coping Strategies Are Created Equal," above, from the questions used in their research to assess why someone was engaging in a distraction.

3. G. Manzanares, G. Brito-da-Silva, and P. G. Gandra, "Voluntary Wheel Run-ning: Patterns and Physiological Effects in Mice," *Brazilian Journal of Medical and Biological Research* 52, no. 1 (2018), https://www.scielo.br/j/bjmbr/a

/kCDDvjgLp5p8gRN3PhJ3jKz/?lang=en.

4. Peng Huang et al., "Voluntary Wheel Running Ameliorates Depression-Like Behaviors and Brain Blood Oxygen Level-Dependent Signals in Chronic Unpredictable Mild Stress Mice," *Behavioural Brain Research* 330 (2017): 17-24, https://www.sciencedirect.com/science/article/pii/S0166432817305363?casa _token=nEqoQfBdjFEAAAAA:zJMS05 https://pubmed.ncbi.nlm.nih.gov /28527694/.

5. Lloyd Demetrius, "Of Mice and Men: When It Comes to Studying Ageing and the Means to Slow It Down, Mice Are Not Just Small Humans," *EMBO Reports* 6, no. S1 (2005): S39-S44, https://www.embopress.org/doi/full/10.1038/sj.embor .7400422.

6. Marc D. Cook et al., "Forced Treadmill Exercise Training Exacerbates Inflammation and Causes Mortality While Voluntary Wheel Training Is Protective in a Mouse Model of Colitis," *Brain, Behavior, and Immunity* 33 (2013): 46-56, https://www.sciencedirect.com/science/article/pii/S0889159113001955 ?casa_token=6m2SH4DquCMAAAAA:ZFtgFRlW5Yu5i8m_SMZMm_6kC6m -xoqGtwAnshO4WVI9nl5eq272ilHgBZ9UQnQOYnvPGjglHw.

7. Carla M. Yuede et al., "Effects of Voluntary and Forced Exercise on Plaque Deposition, Hippocampal Volume, and Behavior in the Tg2576 Mouse Model of Alzheimer's Disease," *Neurobiology of Disease* 35, no. 3 (2009): 426-432, https://www.sciencedirect.com/science/article/abs/pii/S0969996109001405 ?casa_token=JmPtm7Fi6L0AAAAA:XJnFFWgHzUNI5OiL7d5zJoLyBcew U8T5kB9oaEbRIs8KLJJaoxZrmiD92829ozKogMSCa9wUCw. You might be wondering why exercise triggers all these health problems. In mice, at least, stress hormones seem to be part of the story. Mice release stress hormones when they're forced to exercise, stress hormones that they don't release when they exercise at their leisure, and one hypothesis is that these stress hormones exacerbate their illnesses. Martina Svensson et al., "Forced Treadmill Exercise Can Induce Stress and Increase Neuronal Damage in a Mouse Model of Global Cerebral Ischemia," *Neurobiology of Stress* 5 (2016): 8-18, https://www.sciencedirect.com/science /article/pii/S2352289516300200.

8. Yuede et al., "Effectts of Voluntary and Forced Exercise."

9. Lauren A. Leotti and Mauricio R. Delgado, "The Value of Exercising Control over Monetary Gains and Losses," *Psychological Science* 25, no. 2 (2014): 596-604, https://journals.sagepub.com/doi/full/10.1177/0956797613514589?casa _token=cNN9LJsTJroAAAAA:kw1GtAXGGrjodPOvR-o1B_khMJu_SRpb7j XqHbhpKarBnhfnBAHpOTKpsICAr6TbG4nqxyZSA1ld.

10. Frank J. Infurna et al., "Long-Term Antecedents and Outcomes of Perceived Control," *Psychology and Aging* 26, no. 3 (2011): 559, https://www.ncbi.nlm.nih .gov/pmc/articles/PMC3319760/.

11. Francesco Pagnini, Katherine Bercovitz, and Ellen Langer, "Perceived Control and Mindfulness: Implications for Clinical Practice," *Journal of Psychotherapy*

Integration 26, no. 2 (2016): 91, https://www.apa.org/pubs/journals/features/int -int0000035.pdf.

12. Ibid.

13. Milagros Bárez, "Perceived Control and Psychological Distress in Women with Breast Cancer: A Longitudinal Study," *Journal of Behavioral Medicine* 32 (2009): 187-196, https://link.springer.com/article/10.1007/s10865-008-9180-5.

14. Maria T. M. Dijkstra and Astrid C. Homan, "Engaging in Rather Than Disen-gaging from Stress: Effective Coping and Perceived Control," *Frontiers in Psychology* 7 (2016), https://www.frontiersin.org/articles/10.3389/fpsyg.2016.01415 /full.

15. Verena Ly et al., "A Reward-Based Framework of Perceived Control," *Frontiers in Neuroscience* 13 (2019): 65, https://www.frontiersin.org/articles/10.3389 /fnins.2019.00065/full.

16. Kainan S. Wang and Mauricio R. Delgado, "Corticostriatal Circuits Encode the Subjective Value of Perceived Control," *Cerebral Cortex* 29, no. 12 (2019): 5049-5060, https://www.ncbi.nlm.nih.gov/pmc/articles/PMC7049308/.

17. Julian B. Rotter, "Generalized Expectancies for Internal Versus External Control of Reinforcement," *Psychological Monographs: General and Applied* 80, no. 1 (1966): 1-28, https://psycnet.apa.org/doiLanding?doi=10.1037%2Fh0092976.

18. Kainan S. Wang, Madhuri Kashyap, and Mauricio R. Delgado, "The Influence of Contextual Factors on the Subjective Value of Control," *Emotion* 21, no. 4 (2021): 881, https://psycnet.apa.org/manuscript/2020-31605-001.pdf.

19. Craig A. Taswell et al., "Ventral Striatum's Role in Learning from Gains and Losses," *Proceedings of the National Academy of Sciences* 115, no. 52 (2018): E12398-E12406, https://www.pnas.org/doi/full/10.1073/pnas.1809833115.

20. Johanna Drewelies et al., "Perceived Control Across the Second Half of Life: The Role of Physical Health and Social Integration," *Psychology and Aging* 32, no. 1 (2017): 76, https://bpb-us-e2.wpmucdn.com/faculty.sites.uci.edu/dist/4/562/files /2020/03/Drewelies-et-al.-Perc-Control-2017.pdf.

21. For research on the relationship between memory and perceived control, see Frank J. Infurna and Denis Gerstorf, "Linking Perceived Control, Physical Ac-tivity, and Biological Health to Memory Change," *Psychology and Aging* 28, no. 4 (2013): 1147. For research on improved health outcomes, see Frank J. Infurna, Denis Gerstorf, and Steven H. Zarit, "Examining Dynamic Links Between Perceived Control and Health: Longitudinal Evidence for Differential Effects in Midlife and Old Age," *Developmental Psychology* 47, no. 1 (2011): 9. For re-search on longer life span, see Frank J. Infurna, Nilam Ram, and Denis Gerstorf, "Level and Change in Perceived Control Predict 19-Year Mortality: Findings from the Americans' Changing Lives Study," *Developmental Psychology* 49, no. 10 (2013): 1833.

22. Ute Kunzmann, Todd Little, and Jacqui Smith, "Perceiving Control: A Double-

Edged Sword in Old Age," *The Journals of Gerontology Series B: Psychological Sciences and Social Sciences* 57, no. 6 (2002): P484-P491, https://academic.oup.com/psychsocgerontology/article/57/6/P484/669674.

23. Drewelies et al., "Perceived Control Across the Second Half of Life."

24. Suzanne C. Thompson, "Maintaining Perceptions of Control: Finding Perceived Control in Low-Control Circumstances," *Journal of Personality and Social Psychology* 64, no. 2 (1993): 293–304, https://psycnet.apa.org/record/1993-22306 -001.

25. Yi-Yuan Tang, Rongxiang Tang, and Michael I. Posner, "Mindfulness Meditation Improves Emotion Regulation and Reduces Drug Abuse," *Drug and Alcohol Dependence* 163 (2016): S13-S18, https://www.sciencedirect.com/science/article /pii/S0376871616001174.

26. Priyanka Malhotra, "Exercise and Its Impact on Anger Management," *Acta Scientific Medical Sciences* 3, no. 5 (2019): 132-137, https://www.researchgate .net/profile/Priyanka-Malhotra-4/publication/332902331_Exercise_and_its _Impact_on_Anger_Management_Mini_Review/links/5efc1432299bf18816f60 950/Exercise-and-its-Impact-on-Anger-Management-Mini-Review.pdf.

27. Jared B. Torre and Matthew D. Lieberman, "Putting Feelings into Words: Affect Labeling as Implicit Emotion Regulation," *Emotion Review* 10, no. 2 (2018): 116-124, https://journals.sagepub.com/doi/full/10.1177/1754073917742706 ?utm_source=nationaltribune&utm_medium=nationaltribune&utm_campaign =news.

28. Kainan S. Wang and Mauricio R. Delgado, "The Protective Effects of Perceived Control During Repeated Exposure to Aversive Stimuli," *Frontiers in Neuroscience* 15 (2021), https://www.frontiersin.org/articles/10.3389/fnins.2021.625816 /full.

29. Xi Yang et al., "vmPFC Activation During a Stressor Predicts Positive Emotions During Stress Recovery," *Social Cognitive and Affective Neuroscience* 13, no. 3 (2018): 256-268, https://academic.oup.com/scan/article/13/3/256/4867907.

30. Silvia U. Maier and Todd A. Hare, "Higher Heart-Rate Variability Is Associated with Ventromedial Prefrontal Cortex Activity and Increased Resistance to Temptation in Dietary Self-Control Challenges," *Journal of Neuroscience* 37, no. 2 (2017): 446-455, https://www.jneurosci.org/content/jneuro/37/2/446.full.pdf. It's worth noting here that the researchers didn't test whether increasing one's HRV would increase activity in the vmPFC. They simply noted that individuals who had higher HRV also had more activity in the vmPFC. It's unclear which causes which, or if some third variable causes both, but it shouldn't hurt to do some skewed breathing and it could help in multiple ways.

31. Christina B. Young and Robin Nusslock, "Positive Mood Enhances Reward-Related Neural Activity," *Social Cognitive and Affective Neuroscience* 11, no. 6 (2016): 934-944, https://academic.oup.com/scan/article/11/6/934/2223532.

Chapter 13

1. Gesa Berretz et al., "Romantic Partner Embraces Reduce Cortisol Release After Acute Stress Induction in Women but Not in Men," *PLoS ONE* 17, no. 5 (2022): e0266887, https://doi.org/10.1371/journal.pone.0266887.

2. Marleen Van Eck et al., "The Effects of Perceived Stress, Traits, Mood States, and Stressful Daily Events on Salivary Cortisol," *Psychosomatic Medicine* 58, no. 5 (1996): 447-458, https://journals.lww.com/psychosomaticmedicine/abstract /1996/09000/the_effects_of_perceived_stress,_traits,_mood.7.aspx. There isn't a perfect correlation, however, between stressful events and one's cortisol levels. Many factors can make stressful events more or less impactful. If, for example, you feel your life is highly meaningful, stressful events don't affect your cortisol levels nearly as much as they would for someone who feels their life is meaningless. Matias M. Pulopulos and Malgorzata W. Kozusznik, "The Moderating Role of Meaning in Life in the Relationship Between Perceived Stress and Diurnal Cortisol," *Stress* 21, no. 3 (2018): 203-210.

3. Karen M. Grewen et al., "Warm Partner Contact Is Related to Lower Cardiovascular Reactivity," *Behavioral Medicine* 29, no. 3 (2003): 123-130. For more recent research, see Perry M. Pauley, Kory Floyd, and Colin Hesse, "The Stress-Buffering Effects of a Brief Dyadic Interaction Before an Acute Stressor," *Health Communication* 30, no. 7 (2015): 646-659.

4. Valentina Russo, Cristina Ottaviani, and Grazia Fernanda Spitoni, "Affective Touch: A Meta-analysis on Sex Differences," *Neuroscience & Biobehavioral Reviews* 108 (2020): 445-452, https://www.sciencedirect.com/science/article /pii/S0149763418308480?casa_token=S-COle_K4e0AAAAA:u0khVuheveTm Jh6kliBtEU013rBSiNEel6zbm5E6NlTPLlYRdUHG-4iQNjnui2_Cqj-Rflx1cw.

5. Aljoscha Dreisoerner et al., "Self-Soothing Touch and Being Hugged Reduce Cortisol Responses to Stress: A Randomized Controlled Trial on Stress, Physical Touch, and Social Identity," *Comprehensive Psychoneuroendocrinology* 8 (2021): 100091, https://www.sciencedirect.com/science/article/pii /S2666497621000655#bib4.

6. Kory Floyd and Sarah Riforgiate, "Affectionate Communication Received from Spouses Predicts Stress Hormone Levels in Healthy Adults," *Communication Monographs* 75, no. 4 (2008): 351-368, https://www.tandfonline.com/doi/full/10 .1080/03637750802512371.

7. Tanya G. K. Bentley et al., "Breathing Practices for Stress and Anxiety Reduction: Conceptual Framework of Implementation Guidelines Based on a Systematic Review of the Published Literature," *Brain Sciences* 13, no. 12 (2023): 1612, https://www.mdpi.com/2076-3425/13/12/1612.

8. Ibid.

9. Melissa G. Hunt et al., "Positive Effects of Diaphragmatic Breathing on Physiological Stress Reactivity in Varsity Athletes," *Journal of Clinical Sport Psychology* 12, no. 1 (2018): 27-38, https://www.researchgate.net/profile/Melissa-Hunt-2

/publication/323437393_Positive_Effects_of_Diaphragmatic_Breathing_on
_Physiological_Stress_Reactivity_in_Varsity_Athletes/links/5aa162e5aca272
d448b36dfe/Positive-Effects-of-Diaphragmatic-Breathing-on-Physiological-Stress
-Reactivity-in-Varsity-Athletes.pdf?_sg%5B0%5D=started_experiment
_milestone&origin=journalDetail&_rtd=e30%3D.

10. Ibid.

11. Olivia Rogerson et al., "Effectiveness of Stress Management Interventions to
Change Cortisol Levels: A Systematic Review and Meta-analysis," *Psychoneuroendocrinology* 159 (2024): 106415, https://www.sciencedirect.com/science
/article/pii/S0306453023003931.

12. Michaela C. Pascoe et al., "Mindfulness Mediates the Physiological Markers of
Stress: Systematic Review and Meta-analysis," *Journal of Psychiatric Research*
95 (2017): 156-178, https://static1.squarespace.com/static/5dee59a02d0d3203
aa1bbc13/t/5ec68bdb4de9665e0c581c61/1590070247683/Systematic+Review
+%26+Meta+Analysis+of+Mindfulness.pdf.

13. Michaela C. Pascoe, David R. Thompson, and Chantal F. Ski, "Yoga,
Mindfulness-Based Stress Reduction and Stress-Related Physiological Measures:
A Meta-analysis," *Psychoneuroendocrinology* 86 (2017): 152-168, https://www
.sciencedirect.com/science/article/pii/S0306453017300409.

14. Gandhar V. Mandlik et al., "Effect of a Single Session of Yoga and Meditation
on Stress Reactivity: A Systematic Review," *Stress and Health* (2023), https://
onlinelibrary.wiley.com/doi/pdfdirect/10.1002/smi.3324.

15. Jeremy P. Jamieson et al., "Optimizing Stress Responses with Reappraisal and
Mindset Interventions: An Integrated Model," *Anxiety, Stress, and Coping* 31,
no. 3 (2018): 245-261, https://files.eric.ed.gov/fulltext/ED585077.pdf.

16. Jeremy P. Jamieson, Wendy Berry Mendes, and Matthew K. Nock, "Improving
Acute Stress Responses: The Power of Reappraisal," *Current Directions in Psychological Science* 22, no. 1 (2013): 51-56, https://www.psychologytoday.com
/sites/default/files/attachments/126767/arousal-reappraisal-review.pdf.

17. This skier analogy comes from the Jamieson et al. "Optimizing Stress Responses" paper cited above.

18. Jenny J. W. Liu et al., "The Efficacy of Stress Reappraisal Interventions on Stress
Responsivity: A Meta-analysis and Systematic Review of Existing Evidence,"
PLoS One 14, no. 2 (2019): e0212854, https://journals.plos.org/plosone/article
/file?id=10.1371/journal.pone.0212854&type=printable.

19. For research on improved exam performance, see Jeremy P. Jamieson et al.,
"Reappraising Stress Arousal Improves Affective, Neuroendocrine, and Academic Performance Outcomes in Community College Classrooms," *Journal
of Experimental Psychology: General* 151, no. 1 (2022): 197, https://psycnet
.apa.org/manuscript/2021-65684-001.pdf. For research on increased cognitive
flexibility, see Alia J. Crum et al., "The Role of Stress Mindset in Shaping Cog-

nitive, Emotional, and Physiological Responses to Challenging and Threatening Stress," *Anxiety, Stress, and Coping* 30, no. 4 (2017): 379-395, https://emotion .wisc.edu/wp-content/uploads/sites/1353/2022/04/Crum-et-al-2016-The-role-of -stress-mindset-in-shaping-cognitive-emotional-and-physiological-responses-to -challenging-and-threatening-stress.pdf.

20. Crum et al., "The Role of Stress Mindset."

21. Lisa Feldman Barrett, *How Emotions Are Made: The Secret Life of the Brain* (London: Pan Macmillan, 2017), p. 189. Barrett is quoting her daughter's karate teacher, Joe Esposito.

22. Barrett, *How Emotions Are Made.*

23. For original work on how stress reappraisal improves cardiovascular output, see Jeremy P. Jamieson, Matthew K. Nock, and Wendy Berry Mendes, "Mind over Matter: Reappraising Arousal Improves Cardiovascular and Cognitive Responses to Stress," *Journal of Experimental Psychology: General* 141, no. 3 (2012): 417, https://www.ncbi.nlm.nih.gov/pmc/articles/PMC3410434/. For more recent research, see Gavin P. Trotman et al., "Challenge and Threat States: Examining Cardiovascular, Cognitive and Affective Responses to Two Distinct Laboratory Stress Tasks," *International Journal of Psychophysiology* 126 (2018): 42-51, https://www.sciencedirect.com/science/article/pii/S0167876017305433.

24. Jason T. Buhle et al., "Cognitive Reappraisal of Emotion: A Meta-analysis of Human Neuroimaging Studies," *Cerebral Cortex* 24, no. 11 (2014): 2981-2990, https://academic.oup.com/cercor/article/24/11/2981/301871.

25. Chhaye Nene, "10 Extraordinary Uses for Yoghurt (Besides Eating It)," The Healthy, April 20, 2017, https://www.thehealthy.com/food/yogurt-uses/.

26. Marius Hoffmann et al., "Brain Activation to Briefly Presented Emotional Words: Effects of Stimulus Awareness," *Human Brain Mapping* 36, no. 2 (2015): 655-665, https://www.ncbi.nlm.nih.gov/pmc/articles/PMC6869641/.

27. Buhle et al., "Cognitive Reappraisal of Emotion," note on page xvi.

Chapter 14

1. Sheldon Cohen, "Social Relationships and Health," *American Psychologist* 59, no. 8 (2004): 676.

2. For the relationship between social support and heart disease, see Angelo Compare et al., "Social Support, Depression, and Heart Disease: A Ten Year Literature Review," *Frontiers in Psychology* 4 (2013): 384. For the relationship between social support and breast cancer, see Bina Nausheen et al., "Social Support and Cancer Progression: A Systematic Review," *Journal of Psychosomatic Research* 67, no. 5 (2009): 403-415. For research on how lonely people die younger, see Carla M. Perissinotto, Irena Stijacic Cenzer, and Kenneth E. Covinsky, "Loneliness in Older Persons: A Predictor of Functional Decline and

Death," *Archives of Internal Medicine* 172, no. 14 (2012): 1078-1084.

3. Sheldon Cohen et al., "Social Ties and Susceptibility to the Common Cold," *JAMA* 277, no. 24 (1997): 1940-1944.

4. Erica Szkody et al., "Stress-Buffering Role of Social Support During COVID-19," *Family Process* 60, no. 3 (2021): 1002-1015.

5. Niall Bolger and David Amarel, "Effects of Social Support Visibility on Adjustment to Stress: Experimental Evidence," *Journal of Personality and Social Psychology* 92, no. 3 (2007): 458-475.

6. Arie Nadler, Jeffrey D. Fisher, and Shulamit B. Itzhak, "With a Little Help from My Friend: Effect of Single or Multiple Act Aid as a Function of Donor and Task Characteristics," *Journal of Personality and Social Psychology* 44, no. 2 (1983): 310.

7. Emre Selcuk and Anthony D. Ong, "Perceived Partner Responsiveness Moderates the Association Between Received Emotional Support and All-Cause Mortality," *Health Psychology* 32, no. 2 (2013): 231.

8. Brett K. Jakubiak, Brooke C. Feeney, and Rebecca A. Ferrer, "Benefits of Daily Support Visibility Versus Invisibility Across the Adult Life Span," *Journal of Personality and Social Psychology* 118, no. 5 (2020): 1018-1043.

9. For what's now the classic study on how invisible support trumps visible support when one member of a couple is stressed out, see this journal article about people studying for the bar exam: Niall Bolger, Adam Zuckerman, and Ronald C. Kessler, "Invisible Support and Adjustment to Stress," *Journal of Personality and Social Psychology* 79, no. 6 (2000): 953.

10. Jakubiak, Feeney, and Ferrer, "Benefits of Daily Support Visibility."

11. Selcuk and Ong, "Perceived Partner Responsiveness."

12. Brett K. Jakubiak and Brooke C. Feeney, "Affectionate Touch to Promote Relational, Psychological, and Physical Well-Being in Adulthood: A Theoretical Model and Review of the Research," *Personality and Social Psychology Review* 21, no. 3 (2017): 228-252.

13. Mary H. Burleson, Wenda R. Trevathan, and Michael Todd, "In the Mood for Love or Vice Versa? Exploring the Relations Among Sexual Activity, Physical Affection, Affect, and Stress in the Daily Lives of Mid-aged Women," *Archives of Sexual Behavior* 36, no. 3 (2007): 357-368.

14. Rosalba Morese et al., "Social Support Modulates the Neural Correlates Underlying Social Exclusion," *Social Cognitive and Affective Neuroscience* 14, no. 6 (2019): 633-643.

Index

mindset
  continuum, 138
  fixed, 130-132, 136-138, 140, 142, 144
  growth, 131-133, 135-138, 142-144
  how to change your, 138-145
  and how you view mistakes, 132-133, 136-137, 140
  neuroscience of, 135-137
  in stressful situations, 251
  while reading this book, 21
mind-wandering, 33, 38, 120
Miracle Gro analogy, 113
mistakes, making, 129-145
  doctor's, 207-208, 213
  monitoring of, 174
  multitasking and, 29-30
  neuroscience of, 133-137
  and one's mindset, 132-133, 135-137, 140
  optimal number of, 133-134
  as opportunities, 140
  strategies for avoiding, 135-143, 145
moderate-intensity exercise, 99-100, 106, 114
money problems, 217, 229, 232, 234
monkey, 134
mood
  avoidance strategies and, 218-219
  creativity and, 48
  decision-making and, 186-188
  exercise and, 101-102
  naming your, 233
  vmPFC and, 234-238
  See also emotion
morning person, 28
motivation, 61-77
  caffeine and, 71-73
  cold exposure and, 69-71
  compassion training and, 158, 160
  dopamine and, 63-71
  to empathize, 151-154, 157
  goals and, 63
  to help others, 160
  neuroscience of improving, 64-68
  strategies for improving, 63, 68-77
  values affirmation and, 73-75
multitasking, 15, 28-31, 32
music, 68-69, 236

N

names, remembering
  effective strategies for, 117-122, 124-127
  intuitive strategies for, 19-20, 125
Neff, Kristin, 163
neurogenesis, 117, 121
neuroimaging, 15-16, 139
  EEG, 36, 135
  ERP, 135-136, 138
  functional magnetic resonance imaging (fMRI), 84
  positron emission tomography (PET), 71-72
neuroplasticity, 113-114, 139
neuroscience
  of bias, 227, 233-237, 241
  changes in, 14-16, 271
  of control, 223-224, 227, 233-237
  course in, 12-14
  of creativity, 48, 50-59
  of decision-making, 186, 191-195, 197-198
  early days of, 14-16
  of empathy, 101-103, 106, 154-157, 159
  of executive function, 101-103, 106
  of focus, 33, 36-37, 120 136
  of goals, 56, 84-85
  of making mistakes, 133-137
  of memory, 103, 112-117, 119-121
  of mindset, 135-137
  of motivation, 64-68
  of productivity, 82-83
  of placebo effects, 203-204, 209-211
  of social support, 266-267
  of stress, 227, 233-237, 241
neurotransmitters
  adrenaline, 126-127
  dopamine, 50-54, 63-72, 104-105, 203-204, 209, 211-212
  endorphins, 203
  norepinephrine, 102-103
  oxytocin, 210-212
  serotonin, 102-103
neurotrophin, 113
night owl, 28
nocebo effect, 207